Soils

Frontispiece A texture-contrast soil, near crest of long gentle hillslope on Mesozoic sandstone, Sydney region.

Soils

A New Global View

T. R. Paton, G. S. Humphreys, P. B. Mitchell
Macquarie University

Yale University Press
New Haven and London

Published 1995 in the United States by Yale University Press, and in the
United Kingdom by UCL Press Limited. The name of University College
London (UCL) is a registered trade mark used by UCL Press with the
consent of the owner.

Typeset in Plantin and Optima.
Printed and bound by
Biddles Ltd, Guildford and King's Lynn, England.

Listed by the Library of Congress.

International standard book numbers: 0-300-06576-0 (cloth)
0-300-06609-0 (pbk.)

10 9 8 7 6 5 4 3 2 1

Contents

CONTENTS

vi

Acknowledgements

We are grateful to the following individuals and organisations for granting permission to reproduce copyright materials:

Figure 1.1, CSIRO; Table 1.1, *Principles of geochemistry*, John Wiley; Figure 1.4, *Quarterly Review of Biology*, The University of Chicago Press; Figure 2.1, *Clays and Clay Minerals*, Clay and Clay Mineral Society; Figures 3.2, 5.2, 5.3, 8.1, *Search*, Control Publications; Figures 3.3, 8.3, 8.4, 8.5, *Australian Journal of Soil Research*, CSIRO; Figures 4.1 and 9.3, *Journal of Ecology*, Blackwell Scientific; Figures 4.2, 4.3, 4.5, 4.8, 4.9, 4.10, 6.5, 6.6, 6.7, 6.8 and Table 4.2, *Geoderma*, Elsevier; Figures 4.7, 4.13, A1.1, A1.2, A1.3, A1.4, A1.5, A1.6, A1.7, A1.8, A1.9, T. R. Paton, *The formation of soil material*, Chapman & Hall; Figure 4.12, *Catena*, Catena Verlag; Figure 6.1, *Landslides and related phenomena*, Columbia University Press; Figure 6.3, *Overseas geology and minerals resources*, HMSO; Figure 6.4, Louisiana State University Press; Figure 7.1, *Geographical Magazine*, Savedish Enterprises; Figure 8.2, *Australian Journal of Experimental Agriculture and Animal Husbandry*, CSIRO; Figure 8.6, Linnanean Society of New South Wales; Figure 9.1 and 9.2, Table 9.1 and 9.2, *Geological Magazine*, Cambridge University Press; Figures 9.4, 9.5, 9.6, 9.7, 9.9, *Journal of Soil Science*, Blackwell Scientific; Figures 9.8 and 9.10, The Biscuit, Cake, Chocolate and Confectionery Alliance; Figure 9.11, INEAC, SERDAT, Brussels; Figures 10.1, *Soil Science Society of America Journal*, SSA Association; Figures 10.2 and 10.3, *The Geographical Journal*, Royal Geographical Society; Figure 10.4, T. L. Péwé; Figure 10.5, *Glacial and fluvioglacial landforms*, Longman; Figure 11.5, *Mountain Research and Development;* Figure 11.6, The estate of H. S. Gibbs, *New Zealand soils*, Oxford University Press.

Colour plates: P. B. Mitchell (Plates 1–2) and T. R. Paton (Plates 3–8).

List of plates

Introduction

... every discipline, as long as it used the Aristotelian method of definition has remained arrested in a state of empty verbiage and barren scholasticism, and that the degree to which the various sciences have been able to make any progress depended on the degree to which they have been able to get rid of the essentialist method.   *Karl Popper*

Critique of soil zonalism

Over the past 40 years there has been an ever-increasing concentration by pedologists on soil taxonomy, with the main aim being to classify soils, in the form of profiles, for a range of pragmatic purposes. Meanwhile, the understanding of soils in the sense of how they have been formed – i.e. **pedogenesis** – has become increasingly neglected and submerged in an ever more precise morphological system and its associated sea of neologisms. As Peters (1991: 40) has so aptly stated:

> Sciences generate jargon. This terminology is necessary to express complex and novel ideas clearly and succinctly, but terminology often becomes a linguistic barrier around the science, protecting its practitioners from too close a scrutiny by the public or other scientists and providing the camaraderie offered by a common but private language. In weak sciences, jargon substitutes for ideas and theories so that even a hopelessly uninformative field can share some of the trappings of success.

The roots of this pedogenic atrophy can be traced back to the climatic determinism and the associated soil zonalism of the early years of the twentieth century, which will now be discussed together with some of the major problems that arose from zonalism and how, in attempting to solve these problems, the present situation arose.

The concept of soil zonalism has dominated pedology throughout the twentieth century and, although its origin can be traced back to Russian

1

scientists of the late nineteenth century, it has been through the United States Department of Agriculture (USDA), especially C. F. Marbut and his successors, that the greatest impact has been made. One of the earliest and most complete statements about the nexus of ideas that go to make up soil zonalism is contained in the USDA *Yearbook of agriculture* for 1938, *Soils and men*, (USDA 1938: 948–92). According to this, soil is produced over **time** by the action of **climate** and **living organisms** upon **parent material**, as conditioned by local **relief**, which results in the formation of **zonal, intrazonal** or **azonal** soils. The most important characteristics of zonal soils result from the effects of the active factors of climate and living organisms, on well drained parent materials of mixed mineral composition, over a long period of time, whereas in the case of intrazonal soils the dominant factors are passive, either parent material or relief. Azonal soils are closely related to their parent material, with few if any characteristics ascribable to pedogenesis.

The main emphasis in this zonalistic scheme is the operation of the active factors downwards from the surface, which leads to the development of two genetically related **horizons**: the surface, or A horizon, from which fine-grain materials are removed in the process of **eluviation**; and the zone immediately below this, where the eluviated material is deposited to form the **illuvial** or B horizon. Together these two horizons form the **solum**. In suitable conditions, continuation of this eluvial–illuvial process leads to an increasing contrast between the two horizons, which ultimately results in the formation of a coarse-textured residual A horizon over a heavier-textured B horizon: the aptly termed texture-contrast soil.

Despite the fact that texture-contrast soils were generally accepted as forming by illuviation (USDA 1938, Thorp & Smith 1949, McCaleb 1959, Stephens 1962), there was little hard evidence to support this contention, and since 1938 a major task of pedology has been to acquire such evidence. Thus, Brewer (1955), in trying to assess the degree to which clay illuviation was involved in the development of a yellow podzolic soil in southern New South Wales, used zircon as a stable mineral to measure the degree of bedrock alteration and so determined the amount of clay that could have been produced by this process. The results showed that less clay was present in all horizons than could have been formed by alteration. Hence there was no need to invoke clay illuviation to explain the clay maximum at the top of the B horizon. Similar conclusions were reached by Green (1966),using the same techniques in the study of a red-brown earth on similar granodiorite bedrock in northern Queensland. Oertel (1961, 1968, 1974) established that gallium proxies for aluminium in clay minerals, and hence could be used as a marker of clay movement in profiles, as could the concentration ratios of alumina, ferric oxide and titania (Oertel & Giles 1967). In both cases no evidence was found of any significant clay translocation contributing to the clay maximum at the top of the B horizon.

These were indirect investigations of clay movement. A more direct method was provided by the recognition of clay skins, or **cutans**, as resulting from the deposition of illuvial clay (Frei & Cline 1949, Brewer 1964, Cady 1965). Brewer (1968) used this approach to investigate the amount of illuviated clay in ten texture-contrast soils (five red-brown earths and five red podzolics). Total clay was determined by particle-size analysis, and illuviated clay estimated by point-counting on random strips across thin sections. In three of the red-brown earths there was virtually no illuviated clay, whereas in the other two no more than 5–6% of the clay supposedly lost from the A horizon could be accounted for as illuviated clay in the B horizon. In the red podzolics the amount of illuviated clay was greater, but it was distributed down through the subsoil and even into the bedrock, rather than being concentrated in a clay maximum at the top of the B horizon.

A much more fundamental investigation on the use of cutans as a means of quantifying illuviation was made by McKeague et al. (1978, 1980, 1981) and McKeague (1983), who tested whether or not a Bt horizon (a subsoil layer containing ≥1% illuvial features) could be consistently identified, either in the field or in the laboratory. It can be concluded from their investigations that there was so much variation between individual investigators, in deciding what was and what was not an illuvial cutan, that the method did not provide an effective way of quantifying the amount of illuvial clay.

It was obvious from these studies that clay illuviation from A to B horizons was an inadequate general explanation for the formation of texture-contrast soils and that the method for quantifying the most direct evidence for illuviation (cutans) was also not without its problems. Despite this, however, the solum persisted as a central pedological concept, so that whatever other explanations were suggested they needed to operate within an eluvial–illuvial framework, of defined A and B horizons. As long ago as 1949, Simonson (1949) realized that in the case of the red/yellow podzolics of the southeastern USA the A horizon was too thin to have provided the amount of clay that occurred in the B horizon. It was, therefore, proposed that clays were formed from primary minerals in the B and C horizons and were destroyed in the A horizon, so that a clay maximum resulted at the top of the B horizon. This idea was widely accepted and applied as a genetic explanation in many parts of the world even though no evidence had been produced bearing on the different conditions that must prevail in the B horizon and A horizon, if clay formation was to occur in one place and clay destruction in the other. A variant of this model was proposed by Brewer (1955)

3

and Green (1966), who suggested that the products of clay breakdown in the A horizon were removed in solution by drainage along the A/B horizon boundary; once again, however, no direct evidence was offered in support of this contention. Another possibility advanced was that the evidence of clay illuviation, that is the cutans, had been destroyed in those situations where B horizon clays were capable of sufficient seasonal expansion and contraction (Nettleton et al. 1969, Chittleborough & Oades 1979, Chittleborough et al. 1984a,b). However, the evidence used (that of the degree of expansion of the B horizon clays) was indirect and circumstantial and no more convincing than any of the other evidence.

In brief, it can be concluded that none of these ideas concerned with the redistribution or differential destruction of fine-grain materials within an originally homogenous mass offered a reasonable solution to the problem of the formation of texture-contrast soils. Nor did the complementary idea of inheritance from bedded material prior to differential alteration as advanced by Oertel & Giles (1967) and Oertel (1974), for whereas there was some possibility that it might apply in very special circumstances in areas of recent deposition, there was no chance of it being a general solution to the problem, especially away from sites subject to riverine or aeolian activity. This conclusion was strengthened when Chittleborough & Oades (1979), in reinterpreting Oertel's (1974) data, maintained that the topsoil and subsoil of the texture-contrast soil involved had developed from what was originally homogenous material.

Chittleborough & Oades (1980a,b) suggested a new variant of clay illuviation, based on the observation that the clay fraction was apparently bimodal – a coarse fraction of lower mobility and a fine fraction of higher mobility, which were distinguishable both chemically and mineralogically – to demonstrate a relative enrichment of fine clay coinciding with the B horizon maximum. This was applied to river valley terrace sequences in southeastern Australia (Chittleborough et al. 1984a,b,c) on the supposition that the age of the soil increased with terrace height, and that all the soils were created from similar sedimentary materials, so that the soils of the highest terrace had been subject to illuviation for the longest period of time and hence had the greatest texture-contrast development. However, the process of differential illuviation of fine- and coarse-grain clay was questioned even within the papers where it was originally proposed when it was conceded that the profile differences could equally be explained by alteration throughout the profile with a loss of fine clay from the A horizon. Furthermore, Walker & Hutka (1979) in studying soil profiles and sedimentary sequences in southeastern Australia had already established that there was a poor correlation between the distribution of fine (particle-size) and illuviated (point-counted) clays. Later still (Walker & Chittleborough 1986) the importance of clay bimodality was called into doubt when no evidence was found of it in the B horizon clay fraction of 20 texture-contrast soils. Indeed, a better interpretation would seem to be that there had been an enrichment of fine clay towards the top of the B horizon as a result of increased alteration, rather than by illuviation from the A horizon.

A second problem that generated much debate concerned the role of surface erosion and deposition in relation to soil genesis, for even though such processes were excluded from pedological consideration in the zonal model of soil formation, the fact that they operated was undeniable. It was, therefore, necessary from a zonalist point of view to demonstrate that such surface erosion and deposition did not interfere with the processes of pedogenesis. This was attempted in two quite distinct ways; first by Nikiforoff (1949) and then by Butler (1959). Nikiforoff postulated that on gently rolling hillslopes the rate of surface erosion and deposition was in balance with the processes of soil formation, so that, on the upper part of such slopes, the surface of the A horizon was removed by erosion at the same rate as the top of the B horizon changed into A horizon material, for according to Nikiforoff, the A and B horizons had to be maintained at a standard thickness. In other words, in the case of such non-cumulative soils the profile sank into the landscape at the same rate as the surface was eroded. On the lower hillslopes deposition took place at an equally slow rate, such that it equalled the rate at which these sediments became A horizon material. At the same time, to maintain an A horizon of standard thickness an equivalent thickness of basal A horizon was turned into B horizon, so that the profiles of these cumulative soils rose in the landscape in time with the deposition. Over the slope as a whole, there was an equilibrium between erosion/deposition and pedogenesis and a mature or zonal soil developed, which persisted independently of time; it ceased to have a history and from a pedological point of view surface erosion and deposition were finessed, so as not to have any significance. This proposition is still generally accepted as it is an implicit component of soil zonalism, despite the fact that the whole thing depends on a very subtle balance between what are in effect independent variables. In addition, it has never been clearly stated whether it is the non-cumulative or cumulative soil that is zonal. Yet, to consider them both as zonal, given their very different origins, is a paradox that has never been considered.

Butler (1959) also argued for the non-pedological status of surface erosion/deposition, but in a rather different way. From the common occurrence of buried soils he concluded that Nikiforoff's ideas of a balance between continuous erosion and soil formation were inapplicable and that

5

rather than being in a state of equilibrium land surfaces were subject to periodic change, such that at one extreme they were very stable when soil formation could occur and at the other were unstable when erosion and deposition was dominant. As with Nikiforoff, the process of surface erosion and deposition was separated from pedological processes, but in this case it was done by using the time factor, which caused Butler's work to have strong historical implications, deftly avoided by Nikiforoff. Butler's postulated periodicity depended upon the recognition of a recurrent sequence of events, referred to as K cycles and defined as being the time interval covering the formation, by erosion and deposition, of a landscape surface and the development of soils on that surface. K cycles were distinguished from one another by numerical subscripts, so that K_1 was the first cycle back from the present, K_2 the second and so on. Figure I.1 shows how two K cycles would

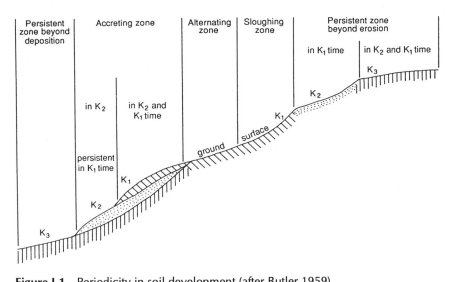

Figure I.1 Periodicity in soil development (after Butler 1959).

have affected a particular hillslope, starting with K_3 as an established ground surface. The K_2 instability brought in erosion on the steeper section of the hillslope and this advanced up slope eroding the K_3 ground surface, while farther down slope the K_3 surface was buried by the debris from this erosion. The K_2 stable phase began when erosion and deposition stopped and soil development began on the K_2 ground surface and advanced farther on the persistent K_3 surface. Later the K_1 instability led to renewed upslope erosion and downslope deposition that was comparable to, but less extensive than, that which occurred in the K_2 cycle. Stabilization of the surface accompanied by renewed soil formation completed the K_1 cycle. It was

6

obvious in this approach that both relic and buried soils were characteristic of these periodic cycles of change, which enabled a history of development to be worked out. The soils developed on the K_1, K_2 and K_3 surfaces were taken to represent a sequence of relic soils of increasing age, such that K_3 soils in the Australian situation were frequently texture-contrast profiles and were viewed as mature or zonal soils, which had developed on a long stable surface.

Despite the fact that erosion and deposition were treated very differently in the Nikiforoff and Butler schemes, pedogenesis was treated in exactly the same way, in that it was restricted to vertically operating processes in both cases. This led to unsustainable, but rather different, endpoints. In Nikiforoff's case the resulting soil was a mature or zonal profile, which once developed was independent of the passage of time. Much the same kind of soil developed on older (K_3) Butlerian surfaces, but instead of being independent of time it was assigned an age of something like 30 000 years, because comparable soil profiles on higher stream terraces had been so-dated. In both cases the resulting soils, which in many cases were texture-contrast, represented a rather sterile endpoint. In Nikiforoff's case the demand was for a balance between two independent variables, erosion/deposition and pedogenesis, over a period of time sufficient for a zonal soil to develop, whereas Butler needed, in the case of K_3 soils, a complete cessation of surface erosion/deposition on a particular landscape segment over a similarly long time period. In view of the efficacy of near-surface processes (see Chs 3–6), neither of these suggestions is really credible.

Other problems also occurred. Thus, the zonalistic introduction of parent material as a factor of soil formation instead of bedrock was also responsible for several difficulties. In *Soils and men* (USDA 1938: 949) it was stated:

> The first step in the development of soil is the formation of parent material, accumulated largely through rock weathering. The parent rock is a relatively inert storehouse of future soil material rather than an active factor in soil formation.

The implementation of such a concept depended on an ability to discriminate consistently between parent material formation and soil formation, and this represented a major difficulty particularly where, as was often the case, both processes were taking place together. It is apparent from the *Soil survey manual* (USDA 1951: 147) that these difficulties were recognized, for it was stated:

> We may conceive of weathering and soil formation as different sets of

processes even though the sets have many individual processes in common and more often than not go on together. Yet a nearly convincing case may be made for considering both together as soil formation, beginning with parent rock as the independent variable in the set of five genetic factors instead of parent material as here defined.

However, having seen all the difficulties and a way of rationalizing them, there was a retreat back to the 1938 concept of parent material.

From this discussion it can be seen that no satisfactory answer has been found to a range of pedological problems. The most important of these concerns the formation of texture-contrast soils. Closely associated is the problem of surface erosion and deposition, where two models have been developed to demonstrate that these processes operate independently of pedogenesis and neither of them have credibility. Another major obstacle was caused by the introduction of the concept of parent material, which led to the exclusion of bedrock alteration from pedogenesis and introduced a large degree of equivocation into pedology. Most particularly it caused increased emphasis to be given to the influence of climate and living organisms (the active factors) at the expense of bedrock and topography, which has had profound consequences on pedogenic understanding.

These problems are all associated with zonalism and its profile-restricted view of pedogenesis, and yet at the present time the zonal system is hardly mentioned except in some introductory texts (Bridges 1978, Gibbs 1980). Indeed, many pedologists would argue that the zonal scheme has been largely superseded by a new morphological system, in which the problems discussed previously are no longer significant. This "new" scheme (*Soil taxonomy*, USDA 1975) maintains that eventually by concentrating on soil morphology and the more refined definition of classificatory units a better pedogenic model will emerge. The newness of approach is most readily seen in the general classification where the three orders of *Soils and men* (USDA 1938), zonal, intrazonal and azonal, are replaced by ten very different orders. However, this change is more apparent than real, for despite the new names all of the orders, except for the inceptisols, can be recognized as having been derived from classificatory units at the suborder and great soil group level in *Soils and men*. Even more significantly differentiation between the orders is still based on maturity as reflected in the degree of horizon contrast and assumed age, and no acknowledgment is made of the great amount of work critical of the concept, which was discussed earlier. In effect the soil orders of *Soil taxonomy* were arrived at by a slight shuffling of the higher classificatory units defined in *Soils and men*, but even more fundamentally

Soil taxonomy remains based on zonal pedogenesis to the same extent as the classic zonal statement given in *Soils and men*. There is, however, a major difference, for whereas in 1938 the emphasis was explicitly on zonal pedogenesis, with classification being derived from it, by 1975 the emphasis was almost entirely classificatory, with zonal pedogenesis being implicit at best. There was a move away from a genetic understanding of soil towards an ability to classify it, a move that had been signalled in the 7th approximation (USDA 1960: 4) where it was stated:

> Since the genesis of any soil is often not understood, or is disputed, it can be used only as a general guide to our thinking in the selection of criteria and forming of concepts. Generally a more or less arbitrary definition of a pedon serves the purpose of classification better at this time than a genetic one.

The continuing belief by a considerable number of pedologists that, by using this approach of more precise definition in soil classification, a better pedogenic model would eventually emerge reflects a belief in **induction**. This means that objective and unbiased conclusions can only be reached by measuring and describing what is encountered without having any prior hypotheses, or preconceived expectations. This viewpoint had been decisively repudiated as long ago as the middle of the nineteenth century, when it was shown to lead to an intellectual impasse and one moreover that no scientist had followed, or ever could follow (Mayr 1982). The most that can be achieved by its application through *Soil taxonomy* (USDA 1975) is an ever-finer splitting of the classificatory units with no possibility of anything novel being generated. *Soil taxonomy* is only the leading example of what has become a general morphological bias. Thus, at a global level there is the FAO–UNESCO (1970–80) *Soil map of the world*, and many countries have developed their own systems. All of these contribute to a general pedogenic atrophy, for all the data involved are constrained by the same zonal criteria and hence have the same limitations. It is as a result of this situation that the authors felt the need for a re-evaluation of pedology, with a return to fundamentals, which means pedogenesis, for without having a firmly based model of soil formation there can be no meaningful classification. Such a reappraisal obviously required a new viewpoint and this came from two sources. First of all there were detailed soil investigations made on a landscape rather than a profile basis, in areas not directly affected by Pleistocene glaciations. Initially this was in Africa on the foundation created by Geoffrey Milne and subsequently in Australia. Supplementing this were a series of process studies by geomorphologists on surface erosion and deposition.

9

This has enabled a new approach to be made regarding the processes of soil formation and the resulting soil materials, which is dealt with in Part I of this book (Chs 1–6).

The second source is plate tectonics, which has so revivified the whole of Earth sciences over the past 30 years, and yet it has been ignored by pedologists, despite the fact that it has completely reformulated our concept of the Earth's surface and in particular how continents are viewed, the very foundation on which pedology rests. In Part II (Chs 7–11) this omission will be rectified and the process/material model of Part I will be expanded on a plate-tectonic base to account for global soil distribution.

PART I

THE PROCESSES OF
SOIL FORMATION AND THE
RESULTING SOIL MATERIALS

Broadly, the processes involved in soil formation belong to one or other of two major groups. One of these deals with the way in which minerals, formed deep within the Earth, react and adjust to near-surface conditions, which can be referred to as **epimorphism**. An understanding of epimorphism can most conveniently start from a consideration of the chemical composition of the Earth's crust, where only eight elements (oxygen, silicon, aluminium, iron, calcium, magnesium, sodium and potassium) account for more than 98% of its weight (Table I.1). Despite being markedly less

Table I.1 The commoner chemical elements of the Earth's crust (after Mason 1966).

	Weight (%)	Atoms (%)	Volume (%)
O	46.60	62.55	93.77
Si	27.72	21.22	0.86
Al	8.13	6.47	0.47
Fe	5.00	1.92	0.43
Mg	2.09	1.84	0.29
Ca	3.63	1.94	1.03
Na	2.83	2.64	1.32
K	2.59	1.42	1.83
Total	98.59	100	100

dense than the other elements, oxygen is by far the dominant constituent and this becomes even more marked when density differences are eliminated, by dividing the weight percentage figures by the element's atomic weight, which shows that out of any 100 crustal atoms nearly 63 are oxygen. If account is taken of the relatively large size of the oxygen atoms (see Fig. A1.5) they will be found to make up nearly 94% by volume of the Earth's crust, which means that the crust can be regarded simply as a packing together of oxygen atoms, with all the other elements accommodated within the packing voids. However, such an accumulation is only made possible by the

11

presence of these other elements, for they are the positively charged cations, which hold together the negatively charged oxygen anions. Because the most abundant cation is silicon, it follows that silicon–oxygen combinations, or silicates, are the most abundant minerals of the Earth's crust, which makes it possible, as a first approximation, to restrict consideration of epimorphic change to silicates and indeed only to those silicates that are of igneous origin, for in terms of the rock cycle everything can be regarded as being initially igneous. Such silicates, which are discussed in Appendix 1, belong to one of four groups – mafic minerals, micas, feldspars and quartz (Table

Table I.2 Silicates.

Tetrahedral linkages	Included elements	Type of silicate	Mineral group
Nil	Fe, Mg	Orthosilicates	Olivine ⎤
Chains	Fe, Mg, Ca, Al	Inosilicates	Pyroxenes ⎥ mafic
			Amphiboles ⎦
Sheets	Fe, Al, Mg, K	Phyllosilicates	Micas
Framework (open)	Al, Ca, Na, K	Tektosilicates	Feldspars
Framework (close-packed)	Nil	Tektosilicates	Quartz

I.2) – all of which are formed at depth within the Earth, under conditions of high temperature and pressure and relatively low oxygen and water concentration. At the Earth's surface, low temperature and pressure, together with abundant oxygen and water, would be expected to change the minerals formed at depth into others more in equilibrium with these surface conditions. Such changes involve:

- **weathering**: the breakdown of primary minerals
- **leaching**: the differential mobility of the breakdown products
- **new mineral formation**: determined by the interaction between weathering and leaching.

In addition account has to be taken of:

- **inheritance**: determined by the amount of material unaffected by or residual from epimorphism. This involves an assessment of the contribution of the bedrock to soil material.

Weathering and leaching deal with the alteration of materials and are considered in Chapter 1, whereas new mineral formation and inheritance are concerned with the products of epimorphism and are discussed in Chapter 2.

The other group of processes stem from the tendency of any topographic variation at the Earth's surface to be eliminated by the action of gravity, which is generally mediated through rainwash, wind and ice. The resulting lateral movement of material across the Earth's surface involves detachment, transport, sorting and deposition, either in particulate form or *en masse*. However, such movement is profoundly affected by the presence of the biosphere, for fauna in particular are responsible for detaching surface material and hence making it more susceptible to surface erosion, whereas the effect of the flora is most particularly seen in its protective role with regard to the processes responsible for lateral movement. Chapter 3 deals with biospheric reactions, then consideration is given to particulate movement attributable to the action of water (Ch. 4) and wind (Ch. 5) before Chapter 6 considers the place occupied by soil creep relative to the other near-surface processes.

CHAPTER 1

Weathering and leaching

The expectations of theory colour perception to such a degree that new notions seldom arise from facts collected under the influence of old pictures of the world. New pictures cast their influence before facts can be seen in different perspective.
Niles Eldredge & Stephen J. Gould

Weathering is that part of epimorphism dealing with the breakdown of minerals. The dominant factor in such breakdown is the hydrogen-ion concentration of the solution in contact with the minerals, for this is responsible for **hydrolysis**, the most important of the processes whereby silicates are weathered. The main source of hydrogen ions results from the reaction between atmospheric carbon dioxide and water to produce carbonic acid, which ionizes to form bicarbonate and hydrogen ions.

$$CO_2 + H_2O = H_2CO_3 = HCO_3^- + H^+$$

The efficacy of hydrolysis is because of the small size of the hydrogen ion, which enables it not only to penetrate the silicate mineral with ease, but also to displace much larger cations from their positions within the lattice. Thus, in the case of feldspars containing large univalent cations such as potassium and sodium, which are coordinated with up to nine oxygens, the individual cation–oxygen bond is weak. The univalent hydrogen ion, because of its small size, can coordinate with only two oxygens between which the single charge is split, which gives rise to a much stronger bond than any associated with potassium or sodium. Thus, the entry of hydrogen ions into a crystal lattice results in the replacement of the larger cations, which will then be subject to leaching. The replacement leads to changes in the feldspar lattice, because the hydrogen is bonded to only two of the oxygens, the remaining seven, previously coordinated with the large cation, have now no such point of attachment and hence will repel one another. The expansion that results increases the strain within the lattice and is ultimately discharged by the breaking of bonds at the crystal surface, with the possible detachment of chains of silicate tetrahedra (de Vore 1959). The rate at which hydrolysis affects the cations in silicate minerals is controlled by the ratio between the valency and the coordination number (Fig. 1.1). The smaller the ratio the greater the ease of hydrolysis; thus potassium hydrolyses easily being univalent and having a coordination number of 9 in feldspars and 12 in micas, giving ratios of 1:9 or 1:12. Aluminium is more difficult to hydrolyse for it is trivalent and has a coordination number of either 6 or 4, which gives a ratio of 1:2 or 3:4. The effect of this variability is reflected in the different reactions of the four main groups of silicates to weathering pressures.

The mafic minerals contain large amounts of magnesium and ferrous iron, which are relatively easily hydrolysable, and their removal leads to complete structural collapse. This has been confirmed by a scanning electron microscope (SEM) investigation of mafic mineral weathering (Berner et al. 1980). The first signs of attack are the development of lens-shaped etch pits in linear series, giving the crystal a striated appearance. The gradual expansion of these pits reduces the crystal to a fragile shell that quickly disintegrates.

The interlayer potassium of micas is easily removed by hydrolysis. However, once this is done, in the case of dioctahedral micas, only silicon and aluminium remain, which are much more resistant to hydrolysis and are more rarely affected, so that for the most part change is restricted to the interlayer zone, whereas the 2:1 silicate framework remains unchanged. The trioctahedral biotites are somewhat different, for their content of ferrous iron and magnesium in the octahedral layer makes them more susceptible to hydrolysis. In certain circumstances complete hydrolysis of these octahedral cations occurs, resulting in the breakdown of the 2:1 phyllosilicate; to all intents and purposes the biotite behaves as a mafic mineral. However,

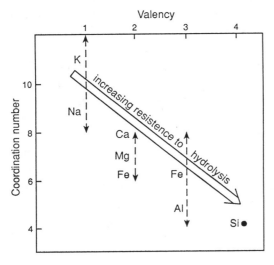

Figure 1.1 Hydrolysis of major cations.

in other situations the ferrous iron is oxidized in place to ferric iron and the resulting excess cationic charge is restored by the expulsion of other octahedral cations, so that a dioctahedral phyllosilicate is formed, which is just as resistant to change as muscovite (Gilkes et al. 1972, 1973b, Gilkes 1973). The rather equivocal situation of the micas is reflected in the work of Jackson et al. (1948) in which clay-size mineral assemblages are arranged in 13 stages of increasing stability, with trioctahedral micas in stage 4 and dioctahedral micas in stages 7 and 8.

In feldspars hydrolysis removes the potassium, sodium and calcium cations, leaving behind a very much weakened framework of tetrahedrally coordinated aluminium and silicon, which gradually disintegrates. SEM investigations (Berner & Holdren 1977, 1979) have confirmed that, as in the case of the mafic minerals, this is initiated through a series of etch pits, the location of which are determined by the outcropping of points of weakness on a particular crystal face; the weathering pattern is determined by the nature of the crystal, not the impinging environment.

In its resistance to hydrolysis quartz stands in stark contrast to all the other silicates, for not only is silicon its only cation, but also the density of packing makes this already unreactive cation (Fig. 1.1) almost inaccessible to the forces of hydrolysis. Thus, under normal conditions of weathering very little happens to quartz apart from a certain amount of size reduction associated with micro-fractures within the grains.

The reactions discussed so far are those that occur under normal conditions, but conditions frequently vary from the normal with considerable effect on mineral stability. Thus, extremely alkaline conditions are generated when alkali and alkaline-earth cations are hydrolysed and, even though these conditions are localized to within a few nanometres (10^{-9} m) of the hydrolysis site, this has considerable importance for the whole process of epimorphism, for at pH values in excess of 9 the solubility of both aluminium and silicon increases markedly (Fig. 1.2; Keller 1957), which has considerable implications not only for weathering but also for leaching and new mineral formation.

The biosphere is also responsible for influencing the pH, but in this case in an acid direction. As a result of the near-surface breakdown of dead organic matter, the carbon dioxide concentration at the biosphere/lithosphere interface can be up to ten times greater than the normal atmospheric level. As this generally occurs in conjunction with a great deal of water adsorbed onto mineral surfaces, it can be concluded that this forms a considerable and continuously replenishable source of hydrogen ions for hydrolysis. Around living plant roots the hydrogen-ion concentration is even greater (Keller & Frederickson 1952), for root surfaces generate a pH

17

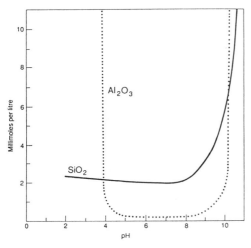

Figure 1.2 Solubility of silica and alumina (after Keller 1957).

in the range of 2–4, which is maintained as long as plant metabolism continues. This is particularly important for aluminium, for below a pH of 4 this normally highly insoluble element becomes very soluble indeed (Fig. 1.2).

It is also possible for the biosphere to participate more directly in the weathering process, for certain organic molecules are able to form complexes with some mineral cations and, in as far as this requires the removal of the cations from the minerals in which they occur, this is a weathering process, called **chelation**. Chelation was demonstrated by Webley et al. (1963), who established that, in the process of colonization of bare rock surfaces, micro-organisms produced oxalic, citric and 2-ketaglucoric acids, which were capable of attacking silicates by forming complexes with their contained cations. Huang & Keller (1972) continued this work by examining the effect of various complexing organic acids on a range of silicates. Their most significant conclusion was that strongly complexing acids extracted up to one thousand times more iron and aluminium from silicate minerals than was removed by a standard distilled water extraction. The removal in particular of so much aluminium from silicates by this chelation process would result in greatly increased framework breakdown, and hence profoundly alter the susceptibility of silicates to weathering. Investigation of rock surfaces colonized by lichens has provided more specific information on these effects. Many lichens excrete oxalic acid in abundance, as well as a variety of weak phenolic acids (lichen acids). It is now evident that, because of its combined complexing and acidic properties, oxalic acid is much more active than either mineral acids or other organic acids and, as it is one of the most abundant acids in soil solution (Vedy & Bruckert 1979), it must have considerable importance in the weathering process. There is

18

still some doubt about the effect of lichen acids, but they do form complexes with cations (Iskandar & Syers 1972) and, given the close association between lichens and outcropping rock, there is no obvious reason why they should not also operate as complexing agents under natural conditions. In fact this is probably the only way they can react, for their weak acidity largely precludes them from acting as acids *per se*.

The introduction of the SEM and microprobe have lessened doubts regarding the weathering ability of lichens, for the more detailed investigation of the rock/lichen interface has revealed the presence of etch patterns beneath a lichen cover, the same as those described previously for feldspars and mafic minerals (Wilson & Jones 1983). The close association between the lichens and the etch patterns makes it certain that the lichens are the responsible agent. The use of the microprobe has enabled the results of the etching to be determined. In the case of a trioctahedral mica in granite covered by lichens (Wilson et al. 1981), magnesium, iron, potassium and aluminium were almost entirely removed, leaving a siliceous relic. In the case of an anorthite feldspar from a more mafic rock, all the calcium and aluminium was removed, leaving behind a siliceous pseudomorph. The ease with which iron and aluminium are removed from these minerals is more consistent with chelation than hydrolysis. Even so a note of warning has been sounded by Mast & Drever (1987) that the effects of chelation in pedogenesis have been somewhat exaggerated. It was contended that under natural conditions the concentration of active molecules, such as those of oxalic acid, may not be sufficient to cause structural breakdown and they were only capable of forming complexes once the cations were freed from the silicate lattice, but were incapable of attacking the lattice itself.

At this point it can be concluded that weathering releases a range of cations – K, Na, Ca, Mg, Fe, Al and Si – together with silicate framework fragments of varying size. The rate at which these ions and framework fragments are produced is related to mineral stability, which varies from the highly unstable mafic minerals to the rather more stable feldspars and micas and ending with the practically inert quartz. At the same time it needs to be recognized that biospheric interactions are capable of profoundly altering mineral stability, although the full extent of these reactions is at present somewhat difficult to gauge.

The degree to which the products of weathering are mobile depends upon their stability in the aqueous environment in which they are released. It is comparatively simple to deal with the silicate framework fragments, for the number of broken bonds that they have would make them very unstable, so that they would not move very far from their point of origin before they react to form new minerals.

19

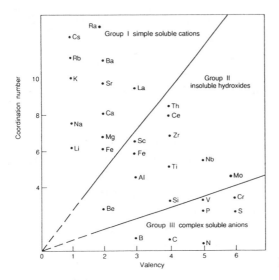

Figure 1.3 Ion mobility groups.

Whether or not cations produced by weathering persist in aqueous solution depends on their reaction with the hydroxyl ion of the water molecule, which, once again, is determined by their ionic potential or the ratio between their size, or coordination number, and valency (Fig. 1.3). If the ratio is small, a simple, soluble cation of group I will result. Intermediate ratios give rise to insoluble hydroxides of group II, whereas elements with larger ratios form complex, soluble anions of group III, such as carbonate or sulphate. It follows from this that in the process of leaching, the elements belonging to groups I and III will be relatively more mobile than the elements of group II, which explains one of the basic characteristics of epimorphism: the general enrichment of residual materials in aluminium and iron. The highly mobile cations of group I are the same as those that are readily freed by weathering (Ca, Mg, Na, K) and for the same reasons: large coordination number and low charge. It is difficult to make a general statement about the relative mobility of ions within this group; in particular circumstances any one of them could be most mobile. In the case of the complex ions of group III, carbonates and sulphates are for the most part only found in soils of arid or semi-arid environments. However, the bicarbonate ion, formed by the ionization of carbonic acid, is of major importance in all environments.

For certain elements, however, this three-way split is not absolute; thus, iron can behave as either a group I or group II element depending on its state of oxidation. If oxygen is lacking it will remain in the ferrous form, unchanged from its usual state in primary igneous rock minerals, and will behave in solution as a stable mobile cation as it does when drainage conditions are poor. However, in the presence of oxygen the highly insoluble ferric form of group II will occur. That this is the common reaction is attested to by the widespread occurrence of red and yellow colours in near-surface rocks and soil materials.

Even though the general solubility of group II elements is low compared to that of the elements making up groups I and III, there is still a considerable range of solubilities within this group. Thus, ferric iron is virtually insoluble no matter what the pH, whereas the solubilities of silicon and aluminium vary with pH, but in rather different ways (Fig. 1.2). Aluminium is highly insoluble within the range of more normal pH values (4.5–9.5); beyond this, in both the highly acid and alkaline environments, solubility increases very sharply. As mentioned previously extreme acidity is generated on root surfaces, whereas extreme alkalinity is associated with hydrolysis, and even though such sites are localized they are epimorphically important, for they are responsible for the mobility of aluminium, which is so essential in new mineral formation, and yet its almost total insolubility at more normal pH values means that it is not moved far from its point of origin. Such a marked and sharp contrast in behaviour depending on the pH is very different from that of silicon, the solubility of which remains constant from alkaline to acid extremes, so that within the normal range of pH (4.5–9.5) silicon is still significantly soluble, which allows a considerable amount of the silicon freed in the breakdown of mafic minerals and feldspars to be removed in solution and end up as one of the important constituents of the dissolved load of rivers and streams.

Biospheric interactions are also important in the process of leaching, for plants generally absorb more of certain elements than others; potassium moves more quickly from the soil into the plant than do other common cations and nitrate more than other anions, so that these two ions are taken up by plants in appreciably higher proportion than their relative abundance in soil suggests should be the case. Hutchinson (1943) investigated this relative accumulation of elements in plants by plotting their concentrations against their concentration in the Earth's crust (Fig. 1.4), which showed that in general the more mobile ions of group I and III are relatively concentrated in terrestrial plants compared to the immobile elements of group II. However, this clear differentiation is somewhat obscured by possible chelation-like processes, which allow group II elements to move and accumulate in plants and organic-rich topsoils. Chenery (1948) in a survey of almost 4000 plants found that almost half of them accumulated aluminium to some extent, whereas iron accumulation was restricted to only one plant. Lovering (1959) showed that in certain cases aluminium accumulation can reach a very high level indeed; thus the Australian tree "white beefwood" (*Orites excelsa*) actually secretes aluminium succinate in cavities within the

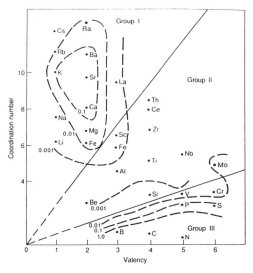

Figure 1.4 Concentration ratio of elements in terrestrial plants and in the crust in relation to ionic potential (after Hutchinson 1943).

tree and its ash is 75% Al_2O_3. Silicon is also accumulated in plants as a result of the formation of opaline silica as phytoliths within cells. This is particularly so in monocotyledons, which can accumulate up to 1% of their dry weight as silica; it also occurs in many trees (Lovering 1959). Where bedrocks of unusual composition occur, it is possible for the vegetation that grows on them to accumulate large quantities of minor elements. Thus, Brooks (1987) lists 57 plants growing on ultramafic rocks that contain in excess of 1% nickel in their leaves and stems. /

In general, then, the biosphere acts as a storage site for elements derived from the Earth's crust, for leaching is retarded, but care should be taken not to overemphasize the importance of such storage, for even where the biomass approaches a maximum, as in tropical forests, the total amount of potassium and phosphorus retained in plants compared to that, say, in a granite bedrock 10 cm thick is only 2–3%, whereas in the case of major elements such as silicon and aluminium the amount stored by the biosphere is very much less (Rodin & Bazilevich 1965).

In overall terms the process of leaching is responsible for the segregation of the more mobile elements (K, Na, Ca and Mg) from the less mobile (Fe and Al) with silicon occupying a somewhat intermediate position. Biospheric reactions, however, can profoundly alter the whole situation.

CHAPTER 2

New mineral formation and inheritance

... "reductionism" is one of those things, like sin, that is only mentioned by people who are against it. *Richard Dawkins*

The nature of new mineral formation in the near-surface zone of the lithosphere is determined by the ambient conditions of temperature and pressure and the flux of materials provided by the processes of weathering and leaching. The low temperature and pressure that prevail in this zone mean that any mineral formed will have a very fine grain size and a disordered structure. It was not until the introduction of X-ray techniques in the 1930s that the crystalline nature of these fine-grain soil materials was established and from then on they were referred to as **clay minerals**. Unfortunately the use of the word "clay" introduced ambiguities, for it had both size (less than $2\,\mu m$) and mineral connotations, and in many cases the newly formed minerals were larger than clay size. Subsequent work established that most of these clay minerals had a phyllosilicate or sheet structure similar to that possessed by the micas. Weaver (1989) suggested that a better term would be **physil**, to refer to this type of phyllosilicate, without any connotation of size, and it is proposed to adopt this suggestion here as a general replacement for the more ambiguous term "clay mineral".

From a general knowledge of silicate structures (Appendix 1) it is apparent that there are two possible routes for physil formation: first, from micas by simple transformation, in which the original sheet structure remains more or less unaltered and, secondly, from mafic minerals and feldspars, which need to be broken down to a much greater extent, before a phyllosilicate structure can be assembled from the fragments.

In the case of micas such as dioctahedral muscovite, the initial epimorphic process is one of physical comminution, which exposes the interlayer potassium to removal by hydrolysis and leaching. This leads to a decrease in bonding strength between the individual 2:1 layers, allowing them to move apart sufficiently to accommodate hydrated cations such as magnesium. The physil formed in this manner is **vermiculite**, which is more reactive than muscovite, for it has a high cation exchange capacity (CEC). A continuation of the physical comminution, and concomitant potassium hydrolysis, decreases still further the bonding strength between the individual 2:1 layers, so that the resulting physil, a **smectite**, acquires a great ability to swell, whereas its CEC is reduced. These changes can only occur as long as there are sufficient hydrated cations such as magnesium and calcium available, which is not the case when strong leaching intervenes and aluminium is left as the only available cation to replace the interlayer potassium. However, when aluminium does this it is in coordination with oxygen and hydroxyl ions as an octahedral sheet, so that the structure changes from 2:1 to 2:1:1 to form a **chlorite** physil. Such chlorites are dioctahedral, as aluminium is the dominant cation in both octahedral sheets and is strongly resistant to further epimorphic change, which is in contrast to the easily weatherable trioctahedral chlorites produced by hydrothermal processes and low-grade metamorphism. When larger mica flakes are affected these changes are mainly confined to the grain periphery, giving rise to zones of chlorite/smectite/vermiculite around a mica core. With smaller particles there is a tendency for interlayers to be affected all across the grain, which gives rise to complexly interstratified materials, a very common characteristic of physils (Fanning et al. 1989).

Feldspars, the most important primary source of physils, present a very different picture from the micas, for their tektosilicate structure needs to break down almost completely before a phyllosilicate can be assembled from the fragments. A complication is provided by many sodic and potassic feldspars being replaced by fine-grain muscovite micas in the late-stage hydrothermal process of sericitization (Harker 1950). In the normal process of epimorphism, however, the non-framework cations, K, Na and Ca, are removed by hydrolysis and leaching, leaving behind a mixture of aluminium and silicon and framework fragments, from which the most common product is a mineral of the kaolin group. Again this has a phyllosilicate structure, but one in which only one tetrahedral sheet is combined with an octahedral sheet to form a 1:1 physil. Kaolins are very inert, for aluminium and silicon are the only cations involved, with the former being confined to the octahedral and the latter to the tetrahedral sheets. However, there is considerable variety in kaolin physils arising from irregularities in the stacking of the 1:1 unit layers, the occupancy of the octahedral sites and the occasional presence of interlayer water. The mineral **kaolinite** is recognized as having an ordered structure and a platy hexagonal habit, whereas **halloysite** represents the disordered end in which interlayer water is often present. It has a variable crystal habit, which may be tubular, spherical or plate-like. With increasing sophistication in the analyses being applied it became apparent that halloysite was the main product of feldspar epimorphism (Parham 1969). Swindale (1975) extended this idea still further by suggesting that halloysite was the only kaolin mineral neoformed in the process of pedogenesis, with kaolinite itself being formed in other ways, hydrothermally or diagenetically. An alternative suggestion was that with time the less ordered halloysite changed to the more ordered kaolinite (Hughes & Brown 1979, Hughes 1980).

Similar ideas have been used in explaining the suite of minerals formed during the epimorphism of volcanic ash and pumice, such as occur in Japan and New Zealand. The feldspars that make up a considerable fraction of these deposits are highly disordered, or even glassy, as a result of their high-speed transfer from magma chamber to lithosphere surface. The disordered structure causes the feldspars to break down quickly when exposed to weathering and leaching and for the resultant materials to recombine just as quickly into the highly disordered aluminosilicates **allophane** and **imogolite**. Despite its amorphous nature allophane characteristically occurs as hollow spheres, 3–5 nm in diameter, with a SiO_2/Al_2O_3 ratio of between 1 and 2 and containing 25–40% water (Wada 1979). There is a greater degree of order in imogolite as it consists of paracrystalline tubes 2 nm across and 1000 nm long. The structure consists of a gibbsite sheet (aluminium in octahedral coordination) bent to form a tube, with SiO_4 groups attached to the inside. Its SiO_2/Al_2O_3 ratio is about 1 and it can absorb 46 g of water/ 100 g clay (Wada 1981). Halloysite was also commonly present and is postulated to form from allophane/imogolite with increasing age. It was further suggested by Fieldes (1955) that halloysite ages to form kaolinite. Some degree of support for this age sequence is to be seen in the work of Tazaki (1986) and Tazaki & Fyfe (1987) on the epimorphism of a highly ordered microcline feldspar. Halloysite was the main physil product, but by using high-resolution transmission electron microscopy (TEM) a poorly ordered highly hydrated silicate with thread-like form, strongly reminiscent of imogolite, was detected in the initial stages of epimorphism. However, such material is ephemeral compared to the long time over which allophane/ imogolite can persist when formed from volcanic ejecta.

Caution is necessary in making such interpretations for there are several other possibilities. Thus, in a volcanic ash deposit, Wada & Matsubara (1968) distinguished allophane, imogolite and gibbsite within a $2 cm^2$ area. The allophane was confined to pumice grains, whereas imogolite only occurred between the grains; gibbsite was restricted to what had been more highly crystalline materials. Such a close association would seem to be better explained by variation of the weathering/leaching balance on a micro-

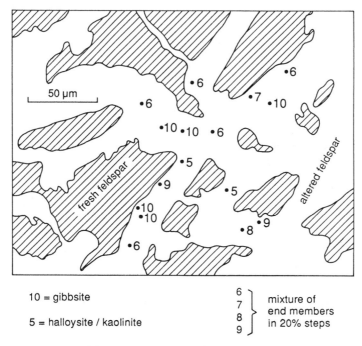

10 = gibbsite

5 = halloysite / kaolinite

6
7
8
9
} mixture of end members in 20% steps

Figure 2.1 Feldspar epimorphism (after Anand et al. 1985).

scale, rather than by invoking a complicated sequential history. The same conclusion would seem to be indicated in the investigation of the epimorphism of individual feldspar grains by Anand et al. (1985) where, within a distance of 50 μm, there are distinct zones of halloysite/kaolinite separated by a narrow boundary from equally distinct zones of gibbsite (Fig. 2.1).

Such conclusions cast doubt on climate being a direct determinant in these processes. These doubts are further strengthened by reports of gibbsite being formed from feldspars from sites as different as a high plateau in Malawi (Young & Stephen 1965), watershed sites on the Maryland Piedmont (Bricker et al. 1968), the highest point of Rhum in western Scotland (Wilson 1969) and the alpine zone of the Cascades in the northwestern USA (Reynolds 1971) where intense weathering and leaching is the only common factor.

The mafic minerals (olivines, pyroxenes and amphiboles) when subject to epimorphism produce a wide range of cations and, if these are not removed by leaching, conditions should be suitable for the formation of a 2:1 physil, as would seem to be the case reported by Eswaran (1979) of a pyroxene being directly replaced by a smectite. However, care is necessary before accepting this example, for smectites and chlorites can be formed from mafic minerals by late-stage alteration (Wilshire 1958, Brown & Stephen 1959, Fawcett 1965), and in view of the widespread distribution of such alteration products within basalts, this must form an important inherited source of smectite-like clays. In addition the plate-like nature of these minerals means that they are easily transported by both wind and water, making them a common contaminant over most of the Earth's surface. In no particular case have all these difficulties been evaluated in a satisfactory manner, so that the question of smectite neogenesis remains in an equivocal state. The normal result of the epimorphism of mafic minerals is for the greater part to be removed in solution, with the residual product being goethite and perhaps hematite (Gilkes et al. 1973a, Eswaran 1979). The structure of these iron oxides can be equated with that of gibbsite, as they consist of octahedral layers, in which the ferric iron is dioctahedral and is coordinated only with oxygen in hematite, but with both oxygen and hydroxyl ions in the case of goethite.

The products of biotite epimorphism can now be considered, for being trioctahedral micas, they lie somewhere between the muscovites and the mafic minerals in their reactivity. Divalent cations, which dominate the octahedral layer in biotite, make it epimorphically more reactive than muscovite. This is reflected in its much easier release of the interlayer potassium. For biotite there are two possible pathways of alteration. In the first, the octahedral ferrous iron is oxidized to ferric and the charge imbalance restored by the expelling of other cations, so that a dioctahedral mica is formed (Gilkes et al. 1972, 1973b). The other pathway is followed if all the octahedral cations are removed, which leads to structural breakdown and concomitant formation of kaolinite, gibbsite and the iron oxides, goethite and hematite, which is the same as happens with the mafic minerals. It has been generally accepted that the first pathway is more characteristic of temperate climates, whereas the second is more usually associated with the tropics (Ojanuga 1973). However, Nettleton et al. (1970), working on a biotite-rich granite in southern California, showed that under good drainage conditions the second pathway was followed, whereas the first pathway was followed under poor drainage conditions. Similarly Eswaran & Heng (1976) showed that in the case of granitic gneiss in Malaysia the same conditions prevailed. Such local environmental controls have been further extended to the micro-scale by the work of Gilkes & Suddiprakarn (1979) on the epimorphism of biotites from a west Australian granite. Within a single biotite grain, oxidation, cation ejection, potassium exchange and the development of secondary minerals occur simultaneously, resulting in the formation of a complex pseudomorph. Electron microprobe analysis shows that such pseudomorphs consist of packets of residual biotite, vermiculite

27

and mixed layer physils, alternating with occurrences of kaolinite, gibbsite, aluminous goethite and hematite. These alternating packets of material, typical of pathway 1 and pathway 2 epimorphism respectively, run parallel to the original biotite foliation, and changes from one to the other occur at right angles to the foliation over a distance of a few micrometres. The two different packets of material reflect rather different weathering/leaching balances, just as was demonstrated in the case of the feldspars. An important additional point is that in this case only plate-like kaolinite is produced; there is no sign of tubular halloysite, despite its abundant production from adjacent feldspars, which indicates that the biotite sheets have controlled the form of the kaolin produced, whereas in the case of the feldspar there is no such rigid control and halloysite is the dominant product.

The question of the biosphere being involved in new mineral formation remains contentious. Fieldes & Swindale (1954) postulated that the concentration of potassium in the soil surface, as a result of plant growth and decay, was sufficient to cause vermiculite to revert to illite by accommodation of the potassium in interlayer positions. Illites found in the surface of Hawaiian soils were ascribed to the same cause (Juang & Uehara 1968). However, it was subsequently shown that the illite was in fact an aeolian addition from the Asiatic mainland (Dymond et al. 1974). Such examples have caused a note of caution to be introduced in invoking possible biospheric effects before more likely alternatives have been evaluated. Just as much equivocation is to be seen in the case of volcanic ash soils. Fieldes (1955) established that in such soils, when humus was abundant, allophane did not form, despite being common in adjacent soils with little or no humus, as a result of the preferential formation of aluminium–humus complexes. However, the evidence on which this conclusion was based was rejected by Kirkman (1975). Subsequently Inoue & Huang (1986) established that a range of organic molecules do form complexes with aluminium and, when this was combined with the more detailed fieldwork of Wada (1985) on Japanese ash soils, which established that the silicon from the aborted allophane formation was confined to humus-rich materials, Fieldes' original contention was re-established. Such swings of opinion are a reflection of the number of unknowns that there are in this subject, for not only is there considerable uncertainty regarding the reactions between the organic and mineral entities, there is also just as much uncertainty regarding the nature of the entities themselves. Thus, although reactions of this kind are probably of considerable importance, much more sophisticated forms of analysis are required to establish what is really taking place.

Therefore, it can be concluded that new mineral formation is controlled by the weathering/leaching balance, which can vary dramatically over extremely short distances, so that different minerals can form concurrently within the confines of a single crystal. This implies that the important control is the micro-environment and not the general climate, which in turn leads to a questioning of the role of climate *per se* in physil neogenesis and soil formation. If it is accepted that the role of macro-climate in soil formation has been overstated it leads to a questioning of the importance attached to the climatic factor in zonalism, which has dominated pedological thought for so long.

Differentiation between inherited and epimorphic products is straightforward as long as the only things inherited are primary igneous rock minerals. However, difficulties are much greater when, as described above, clay minerals are formed by the late-stage hydrothermal alteration of primary igneous rock minerals, for in any subsequent period of pedogenesis they tend to resist epimorphic change and persist. In other words they are inherited, and failure to differentiate them from the very similar epimorphic physils can lead to erroneous pedogenic interpretations. This was the case when a pedogenic evaluation was made of closely juxtaposed red and black clay soils on the Darling Downs in southern Queensland, 150 km west of Brisbane (Beckmann et al. 1974).

The marked soil difference was explained in terms of site, with red soils on flat hilltops and black soils on the adjacent gently sloping pediments. It was argued that the shape of the surface at any site largely determined the amount of runoff and hence the amount of water available for penetration into the soil. On the flat-topped hill-crest through-drainage was good and hence the red soil developed, rich in kaolin and hematite. Because of its flocculated nature through-drainage was maintained and hence the red soil continued to form. In contrast, on the gently sloping pediments, runoff was high and water penetration restricted, so that alteration did not proceed beyond the smectite stage. In addition the presence of these clays increased the degree of runoff, for they slaked and sealed on wetting, effectively preventing any water penetration into and percolation through these soils. Erosion by the high runoff removed the highly altered surface and preserved the soil material at the smectite stage over a long period. Thus, the presence of such closely juxtaposed yet different soils was explained solely in terms of epimorphism, albeit somewhat contrived.

However, the presence of two anomalous red soils on gentle pediment slopes was explained in terms of variations in the nature of the basaltic bedrock (Beckmann et al. 1974). An investigation of the nature of the bedrock (Paton 1974) showed that the flat hill-crests were formed of hard, fresh basalt, whereas the pediments were cut across a series of less resistant lava flows, many of which had been altered to a smectite / calcium carbonate mix, typical of late-stage hydrothermal alteration (Krismannsdòttir 1982, Mehegan et al. 1982, Viereck et al. 1982). The nature of the bedrock controlled both the sites and the style of pedogenesis. The fresh basalt, resistant to erosion, formed the flat hill-crest, which was responsible for good through-drainage and the neoformation of kaolin and hematite, which agrees with the interpretation of Beckmann et al. (1974). This was not the case for the hydrothermally altered basalt, which was considerably less resistant and eroded to form the gentle pediment, whereas the smectite was inherited from the bedrock to form the soil. The hydrothermal inheritance also explained the presence of the abundant calcium carbonate, which could not be explained in terms of epimorphism.

Up to this point consideration has been given to epimorphism and inheritance in terms of *in situ* igneous rocks. The approach can now be widened to follow the end-products of epimorphism – epimorphic 2:1 and 1:1 physils and inherited sand-size quartz – as they are moved across the surface of the Earth, from their point of origin. As a result of such movement they are sorted and then deposited as part of a sedimentary body, a process that may be repeated many times. In all cases of deposition the material is gradually buried and subject to increased temperature and pressure; in other words the materials are subject to diagenesis and low-grade metamorphism, the effects of which differ according to the nature of the materials involved.

In the case of the 2:1 physils the epimorphic sequence of muscovite, vermiculite, smectite, chlorite tends to be reversed in the direction of muscovite. However, a total reversal is not possible because the temperatures and pressures reached are too low and there is not sufficient potassium available. Instead illite is formed with less potassium and more water than muscovite (Norrish & Pickering 1983, Weaver 1989, Lanson & Champion 1991). Illite commonly forms the bulk of many shale deposits, where it is interlayered with smectite and referred to as the I/S complex.

The behaviour of 1:1 or kaolin-type physils is rather different for with deep burial they disappear. The depth and temperature at which this happens vary widely and the factors that determine the reaction are not well known (Weaver 1989). Near-surface diagenesis may, however, contribute to the accumulation of kaolinite. Earlier in this chapter the formation of allophane, imogolite and halloysite was discussed in terms of a possible age sequence, which ended in highly ordered kaolinite. An alternative is that the sequence results from increasing diagenesis with burial (Weaver 1989).

Sand-size quartz is the commonest and bulkiest product of most *in situ* acid igneous rock epimorphism because it is so inert. For the same reason little can happen to it after transport and burial beyond being lithified into a quartz sandstone.

When these materials are once again exposed to the processes of epimorphism, reactions are limited. In the case of the illite/smectite complex (I/S), smectite is formed at the expense of the illite, so that illite can be regarded as being involved in a reversible reaction with smectite, with the balance between the two being controlled by the relative strength of epimorphism and diagenesis. Kaolinite and quartz are both so inert that little or nothing happens when they are subject to epimorphism. It follows that in soils derived from such materials the degree of mineral inheritance must be very high indeed. Cases such as this where epimorphic reactions are very limited and the degree of inheritance is very high are examples of **secondary epimorphism**. They can be contrasted with the **primary epimorphism** of igneous rock minerals where weathering and physil formation are high and inheritance low.

The situation changes considerably when bedrock has been subject to higher grades of metamorphism. In these circumstances new minerals are formed such as micas and feldspars and their subsequent involvement in epimorphism is of course of a primary rather than a secondary nature. Thus, the more intense the metamorphism to which a rock has been subject, the more will any subsequent episode of epimorphism be primary rather than secondary.

The global importance of inheritance/secondary epimorphism can be gauged from the fact that over 70% of the Earth's land surface is covered by unconsolidated and lithified sedimentary materials. This means that over most of the Earth's land surface there is a largely inert cover. Therefore, despite the emphasis that is normally given to primary epimorphism, the dominating role of secondary epimorphism/inheritance needs to be recognized.

Up to this point inheritance has been discussed in terms of mineralogy only. However, there is another type that is just as important and this is **fabric inheritance**, where the spatial arrangement of the constituent minerals of the bedrock are preserved in the soil material, even though the minerals themselves may have been changed by epimorphism. This covers a very wide range of possibilities for it includes everything described as fabrics and structures in igneous, metamorphic and sedimentary rocks, which can be recognized as persisting in some form in soil material and varying in scale from bedding planes and joints down to the micro-cracks that develop in granitic quartz grains. Materials in which there is such fabric inheritance are called **saprolite** (Eswaran & Bin 1978, Hart et al. 1985, Pavich 1989). In addition terms such as **ghost** or **relict** are widely used in this context.

CHAPTER 3

Bioturbation

Innocent, unbiased observation is a myth. *Peter Medawar*

Bioturbation is concerned with reactions between animals, plants and soil material, during which soil fabric is altered by detachment, transport, sorting and deposition of material, both within the soil mantle and on its surface. These processes have been variously discussed by Paton (1978), Hole (1981), Yair & Rutin (1981), Humphreys & Mitchell (1983), Goudie (1988), Mitchell (1988), Johnson (1990), Lobry de Bruyn & Conacher (1990) and Humphreys (1994a), and will now be examined in some detail, first of all with respect to invertebrates and then the vertebrates, before the effects of plants are considered.

Invertebrates

Earthworms
There are some 1800 known species of earthworm (Edwards & Lofty 1977), many with a cosmopolitan distribution, ranging from areas with permafrost to the tropics, but with a preference for humid or seasonally humid areas (see Table 3.1 for a classification of environments).

Darwin (1881) initiated work on earthworm–soil interactions in seeking to explain the formation of the ubiquitous fine-grain surface "vegetable mould" in terms of earthworm casting. This he did by extending his observations from the actual nature and rate of surface casting by earthworms, to the rate of burial of marker horizons, such as lime and cinders, which were periodically spread across the surface, and then to the depth of burial of Roman ruins beneath the "vegetable mould". He showed that disturbance by earthworms extended to a depth of 2.5m and that all the "vegetable mould", which was solely the product of earthworm casting, was in such a fine-grain state and in such a topographic situation that it could easily be moved by the action of wind or rain.

Renewed interest in earthworm–soil interactions arose in the 1940s,

Table 3.1 Environmental categories (after Young & Saunders 1986).

Environment	Symbol	Köppen class	Approximate description
Polar/montane	P/M	E	Periglacial high-latitude and mid-latitude montane areas
Temperate maritime	Tm	Cfb,c	Includes western Europe and east coast USA
Temperate continental	Tc	D	Includes eastern and central Europe; humid interior of USA; northern Asia
Mediterranean	Med	Cs	Includes similar climates in other continents, e.g. California, and southern and southwest Australia
Semi-arid	SA	BS	Approx.250–500mm rainfall zone, e.g. central Australia, western USA, central Asia
Humid subtropics	ST	Cfa	Includes southeast Australia, and USA
Tropical wet and dry	TrS	Aw	Savanna areas, e.g. northern Australia, south and east Asia
Humid tropics	TrH	Af,Am	Permanently wet tropics, e.g. equatorial Southeast Asia, Africa and South America
Arid	A	Bw	<250mm of rain approx, e.g. northern Africa

when it was established that not all earthworms cast on the surface. Evans & Guild (1947) showed that only two out of six species commonly present in Rothamsted pastures were important surface casters. Satchell (1958) established that there were only three surface casters out of 28 species common in the UK. This work showed that earthworm soil bioturbation rates could not be based on the rate of surface casting, but that rather more elaborate methods needed to be developed. A first attempt to solve this problem was made by Evans (1948), who, after showing that surface casts were mostly the work of *Allolobophora longa* and *A. nocturna*, assumed, as a first approximation, that all the soil ingested by these two species ended up as surface casts and that, weight for weight, non-surface–casting species consumed an equal amount of soil, but voided it beneath the surface. Knowing the total weight of the worm population it was then a simple matter to calculate the weight of soil ingested by the non-surface–casting species. Using this approach it was estimated that, for an old pasture at Rothamsted, the weight of earthworm casts varied from 250 to 6250 g m^{-2} yr^{-1} and that subsoil voiding accounted for 500 to 5250 g m^{-2} yr^{-1}. It was realized that these figures were minimum values since not all the soil ingested by *A. longa* and *A. nocturna* ended as surface casts. In addition, in many cases the potassium permanganate vermifuge treatment used to estimate the earthworm population was not efficient, for it penetrated the soil unevenly and earthworms responded to it in a variable way. Nevertheless, these minimal figures were

34

enough to suggest that a square metre of soil to a depth of 10 cm would be totally ingested by earthworms in a period of 15–20 years.

In a more natural (i.e. non-agricultural situation), Satchell (1967), working on some woodland soils in the English Lake District, attempted to solve the quantification problem in a different way. It was shown that on average individuals of *Lumbricus terrestris* contained 100 mg of soil for every gram of their body weight and that soil took a day to pass through the earthworm's gut. If this were maintained for 200 days in the year, the effective plant growing season, by an earthworm population with a mean biomass of 120 g m^{-2}, 2400 g m^{-2} yr^{-1} of soil would be ingested, giving a turnover time for the surface 10 cm of 70 years.

More detailed work on surface casts by Satchell (1958) showed that the earthworms he was dealing with did not ingest particles greater than 2 mm in diameter, so that finer-grain surface casts were produced. Darwin (1881) was aware of this fining effect, but he wrongly ascribed it to the attrition of the ingested particles in the earthworm's gizzard. Parle (1963) showed that the water-stable nature of earthworm casts when first produced was the result of the growth of fungal hyphae, but this disappeared within a few weeks with the death of the fungus, leaving behind a fine-grain material very susceptible, as Darwin had pointed out, to erosion by wind and water.

Nye (1954, 1955a,b,c), from work in the coastal forest zone of southern Nigeria, established that a finer-grain surface layer was produced by earthworms casting at the rate of 5100 g m^{-2} yr^{-1}. Under the somewhat drier conditions of the savanna zone of the Ivory Coast, Lavelle (1978) determined the earthworm casting rate to be 2180–2780 g m^{-2} yr^{-1} contributed by only 6 of the 18 species present, which, it was estimated, accounted for only 4% of the total soil ingested. At this rate of ingestion soil to a depth of 1 m would be turned over in less than 14 years. However, no data were available in this case of a possible finer-grain surface layer.

Recently a certain amount of data with a pedological emphasis have come from work on earthworms in southeastern Australia (Humphreys 1985, Mitchell 1985). From the sites investigated it appeared that there was an extreme contrast in the way that earthworms react with soil on either side of the boundary between subalpine woodland and wet sclerophyll forest. Above the height of this boundary in the Snowy Mountains (Fig. 3.1), that is, above 1100 m, there was an accumulation of mull-like material (Kubiena 1953, Handley 1954) that consisted of organic loam casts derived from the mixing of mineral soil material with well humified organic matter. These casts were highly water-absorbent but retained stability. As a result these soils had an extremely high water retention rate and, even with rainfall intensities estimated at 100 mm h^{-1}, no surface flow was observed. Such

35

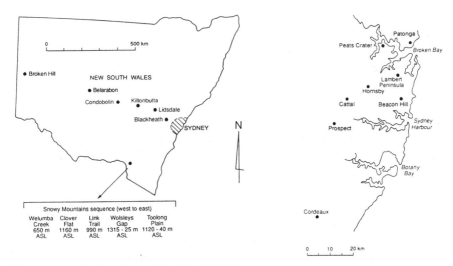

Figure 3.1 Site locations, New South Wales.

fabric stability and lack of surface flow meant that processes of fabric break-down, sorting, transport and deposition were curtailed.

At elevations below 1100 m earthworms were not the dominant bio-turbators. The earthworm casts were neither so absorbent of water nor so stable, presumably because organic matter was less abundant. At Cordeaux, a near-coastal site on Sydney Basin sandstone (Fig. 3.1), earthworm casting amounted to $133 \, \mathrm{g \, m^{-2} \, yr^{-1}}$ with activity concentrated in the winter months (Fig. 3.2; Humphreys 1981). The rate of casting was very significant, for susceptibility of casts to rainsplash breakdown and the relative concentration within them of rather finer material meant that they could make a considerable contribution to the suspended load, which was a feature of overland flow on these slopes.

These investigations in eastern New South Wales established that earth-worm activity was pedogenically important, even in areas previously regarded as being inhospitable to them. In the Snowy Mountains, where the importance of earthworms had previously been noted and they were found to assume a dominant role in pedogenesis, a totally different suite of near-surface processes operated, which resulted in a totally different soil material. Whereas in Australia the Snowy Mountains type of environment is very restricted, it appears that in New Zealand and even more so in Europe and North America comparable environments are more extensive, so that a sim-ilar earthworm-determined soil fabric could be very common. An equiva-lent situation is found in the montane forests in New Guinea, especially on volcanic ash where earthworm activity, in combination with abundant

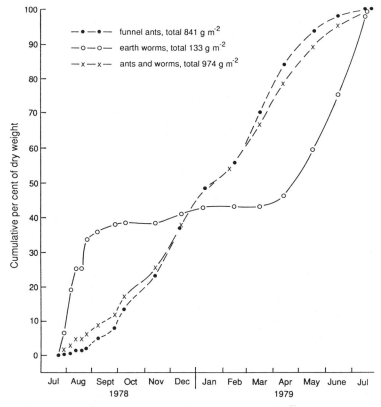

Figure 3.2 Earthworms and ants, rate of surface casting (after Humphreys 1981).

organic matter and allophanic minerals, yields stable aggregates.

To assess the total amount of bioturbation resulting from earthworm activity means taking account of both their subsoil and surface actions. The latter is the comparatively simple task of determining the rate of surface casting; even so there are difficulties resulting from erosive loss between collections, or the possibility that collections might stimulate worms to produce more casts. However, this is straightforward compared to assessing subsoil bioturbation. It can, however, be determined where the rate of soil ingestion by earthworms is known, but again such data are available in only a few cases (Satchell 1967, Lavelle 1978). It follows, therefore, that the bioturbational activity of earthworms is generally recorded in the form of rate of surface casting only (see Table 3.2), which must be taken as a minimum estimate. Over half the studies record rates of 10–50 t ha^{-1} yr^{-1}, especially in the seasonally wet/dry tropics and in temperate latitudes, with lower rates in drier and montane settings. In several cases the rates exceed 100 t ha^{-1} yr^{-1}, for example in Bangalore, India (Krishnamoorthy 1985), northeast Thai-

Table 3.2 Rate of casting by earthworms.

Location	Env. code*	Species	Rate of mounding (t ha^{-1} yr^{-1})	References	Notes
Abidjan, Ivory Coast	TrS	Not identified	10–50	Roose (1980)	Under forest
Ibadan, Nigeria	TrS	Hippopera nigeriae	51.2	Nye (1955c)	Under forest
Lagos	TrS?	Siphomogaster sp.	244	Nye (1955c)	
Cameroons	TrS?	Not identified	70–210	Kollmannsberger (1956, in Edwards & Lofty 1977)	Montane savanna
Gezira, Sudan	?	Not identified	268	Beauge (1912, in Evans & Guild 1947)	
Ibadan, Nigeria	TrS	Hyperiodrilus africanus, Eudrilis eugeniae	175–225	Madge (1965, 1969)	On non-tilled plots under various crops the rate of casting over 70 days varied from 1.1 to 47.6 t ha^{-1} (Lal 1976)
Lamto, Ivory Coast	TrS	Millsonia anomala, M. lamtoiana, Dichogaster agilis, D. terrae-nigrae, Chuniodrilus zielae, Stuhlmannia porife	21.8–27.8	Lavelle (1978)	Lavelle estimated the rate of ingestion (including casting) as 730 to 1100 t ha^{-1} yr^{-1}
Cape District, South Africa	Med	Microchaetus sp.	24–50	Ljungström & Reinecke (1969)	nb: the rate in the micro-landform type (kommetjies) is up to 270 t ha^{-1} yr^{-1}.
England	Tm	Not identified	18.8–40.4	Darwin (1881)	Guild (1955) gives a general figure of 27 t ha^{-1} y^{-1} for English pasture. Satchell (1967) recorded the rate of ingestion by Lumbricus terrestris in the Lake District, England, as 26.4 t ha^{-1} y^{-1}.

Table 3.2 Rate of casting by earthworms.

Location		Species	Rate	Reference	Notes
Rothamsted, England	Tm	Allolobophora longa, A. nocturna	27.6–28.9	Evans & Guild (1947)	Evans (1948) for the same site and species gives a wider range of 5.77 to 61.76 t ha^{-1} yr^{-1}.
Ardennes, Luxembourg	Tm	Allolobophora nocturna	15	Hazelhoff et al. (1981)	
Zurich, Switzerland	Tc	Not identified	9–81.4	Stockli (1929, quoted in Evans & Guild 1947)	
Bellinchen, Germany	Tc	Not identified	5.39–6.8	Kollmannsberger (1934, quoted in Evans & Guild 1947)	
Wroclaw, Poland	Tc	Not identified	8.8–91.6	Dreidax (1931, quoted in Evans & Guild 1947)	
India	TrS	Not identified	1.4–77.8	Roy (1957, in Edwards & Lofty 1977)	Under grassland
Orissa, India	TrS	Lampito mauritii and other species	77.9	Dash & Patra (1979)	Under grassland
Bangalore, India	TrS	Lampito mauritii dominant	102	Krishnamoorthy (1985)	Under forest
Bangalore, India	TrS	Pheretima elongata dominant	137	Krishnamoorthy (1985)	Under grassland
Nam Phong, Thailand	TrS	Pheretima sp.	132.6–224.9	Watanabe & Ruays-oongnern (1984)	Under grassland

Table 3.2 Rate of casting by earthworms.

Location		Species	Rate	Reference	Notes
Kyoto, Japan	ST	Pheretima hupeiensis	38.3	Watanabe (1975)	Hata (1931, in Watanabe & Ruaysoongnern 1984) recorded 281.7 t ha^{-1} yr^{-1} in Japan
Wisconsin, USA	Tc	Allolobophora caliginosa, Lumbricus terrestris	40.4	Neilsen & Hole (1964)	
Richmond, Indiana, USA	ST	Not identified	12.1	Thorp (1949)	
Pico del Oueste, Puerto Rico	TrH	Not identified	1–2	Lyford (1969)	Montane forest
Chimbu, PNG	TrH	Amynthas corticis, Pontoscolex corethrurus	2.3–6.2	Humphreys (1984)	Exotic spp. under pasture
Adelaide, SA, Australia	Med	Allolobophora caliginosa, Eisenia rosea (exotic)	2.5	Barley (1959)	nb. Barley estimated the rate of ingestion (including casting) as 30 to 40 t ha^{-1} yr^{-1}. Exotic spp.
Killonbutta, NSW, Australia	ST/TC	Heteroporodrilus sp.	0.063	Mitchell (1985, 1988)	Under woodland
Cordeaux, NSW, Australia	ST	Cryptodrilis sp. Oreoscolex sp.	1.33	Humphreys (1981, 1985)	Dry sclerophyll forest
Snowy Mtns, Aus.	Tm	Vesiculodrilus purpureus	9.4	Mitchell (1985)	Wet sclerophyll forest
Snowy Mtns, Aus.	M	Vesiculodrilus purpureus	45.4	Mitchell (1985)	Subalpine woodland
New Zealand	Tm	Not identified	2.51	Evans (1948)	
Palmerston North, New Zealand	Tm	Allolobophora caliginosa (exotic)	25–30	Sharpley & Syers (1976, 1977)	Exotic spp.

* See Table 3.1

land (Watanabe & Ruaysoongnern 1984) and West Africa (Madge 1969). These high rates support earlier claims from the Lagos region by Millson (1830, quoted in Nye 1955c), in the Gezira of the Sudan by Beauge (1912, quoted in Evans & Guild 1947) and in Cameroon by Kollmannsberger (1956, quoted in Edwards & Lofty 1977).

Ants

There are some 15 000 known species of ants distributed from arctic regions to the tropics and, unlike earthworms, they are also important in dry regions. They are social insects that live in large colonies and many of them burrow extensively, generally within 2 m of the surface, in the process of constructing nests, which sometimes protrude as mounds on the surface. From the earliest times it was these mounds that attracted most attention (Shaler 1891, Branner 1896, 1900, 1910) and generally they have been used as a basis for calculating the amount of ant bioturbation. The difficulty with this is that ants' do not invariably produce mounds, for by far the greatest number of ants nests are excavated in soil under stones and logs without any recognizable superstructure.

An example from the natural bushland in the Sydney region of eastern Australia is the sugar ant (*Camponotus consobrinus*), which often leaves no sign on the surface despite considerable subterranean activity (Humphreys 1985). The propensity to build mounds varies in other ways also: many species only construct mounds when the nest is well established, for during the early stages of colony development the nest is entirely subterranean and the excavated pellets of soil are carried some distance from the opening and scattered irregularly over the surface. In southeastern Australia some of the bulldog ants (*Myrmecia* sp.) appear to operate in this manner. In other situations the presence of mounds is influenced by habitat conditions. Thus, in western Europe *Lasius flavus* commonly builds mounds, except on lighter-textured and better-drained soils, when the whole nest system is subsurface (Waloff & Blackith 1962).

Although mounds vary widely in size, shape and composition (Wheeler 1910) there is, from a pedological viewpoint, a need to distinguish between two general categories only: type-I and type-II mounds (Humphreys & Mitchell 1983). In type-I mounds, material is simply deposited on the surface, in the same way as common earthworm casts, and the mounds are similarly very susceptible to erosion. In contrast, type-II mounds are larger, compact, often cemented, and form an integral part of the nest. They also persist for a much longer time, for they are markedly resistant to erosive processes and when damaged they are repaired quickly.

The funnel ant (*Aphaenogaster longiceps*), a nocturnal insectivore, was

41

recognized as a very important bioturbator that produced a type-I mound (Humphreys & Mitchell 1983). Mounds were crater-shaped and varied in size from 5 cm to 25 cm in diameter and from 0.5 cm to 11 cm in height (Plate 1). The mean size of single mounds was 8 cm in diameter (across the funnel rim), 2.4 cm high with side slopes of 30–35° and a base diameter of 15 cm. Such a mound had a mean volume of 240 cm^3 and contained 190 g of mineral soil. Mounds were often built beneath fallen logs, against stones and around the base of trees, and consisted entirely of mineral soil piled as single grains of sand, or as very fragile aggregates of up to four grains of fine sand linked by clay bridges. In comparison with the surrounding surface soil the mound was depleted in gravel, coarser sand and clay and often showed internal layering, the layers being demarcated by rainsplashed surfaces and litter, demonstrating that formation was periodic. The central funnel of each mound tapered from 2 cm across at the surface to 1 cm at a depth of 3–8 cm, where it became an enlarged chamber from which several horizontal galleries extended to connect adjacent mounds and shafts. These galleries were ovoid in cross section, 4–8 mm high and 7–20 mm wide, often running parallel to large plant roots that acted as roof to the gallery. The mound density at Cordeaux (Fig. 3.1) averaged 162 per 100 m^2, with a late summer maximum of 379 per 100 m^2. The rate of mound construction was 841 g m^{-2} yr^{-1}, with a summer maximum, which complemented the winter maximum of earthworm casting (Fig. 3.2; Humphreys 1981). In comparison the mound density at Killonbutta (Fig. 3.1) was about 150 per 100 m^2, but at any given time 30–40% of them were inactive as judged from their rainsplashed or litter-covered surfaces (Mitchell 1985). At this site the rate of mound construction on the better-drained upper slope was 47.4 g m^{-2} yr^{-1}, whereas on the more poorly drained lower slopes it was only 8.5 g m^{-2} yr^{-1}. The figures were very much a minimum value, for the mounds were very susceptible to erosion and a great deal was undoubtedly lost during the month or so between collections.

Figure 3.3 is a cross section of a typical type-II ant mound (*Camponotus intrepidus*; Cowan et al. 1985) as developed in the Sydney region of eastern Australia. The mounds were on average 20 cm high, elliptical in shape, and covered an area up to 1.2 m^2. They were formed around clumps of grass or sedge and covered by a thatch of leaves, twigs and charcoal fragments worked into the surface. The mounds formed the top of nests, which in places extended down to almost a metre below the surface and had total volumes that varied from 3000 to 11 000 cm^3. The mound, and the topsoil beneath it, consisted of thoroughly reworked soil material made up of 5–8 mm clumps of subsoil clay, intermixed with topsoil sand and coarse organic debris. Overall this material showed an enrichment in silt and clay

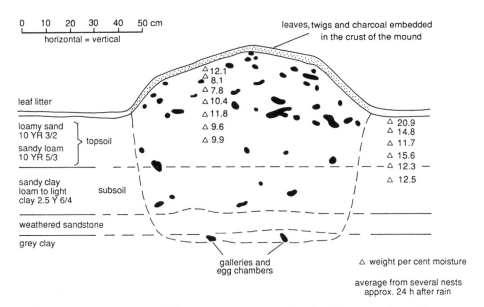

0   10   20   30   40   50 cm

horizontal = vertical

leaves, twigs and charcoal embedded
in the crust of the mound

leaf litter

loamy sand
10 YR 3/2

sandy loam
10 YR 5/3

topsoil

△ 12.1
△ 8.1
△ 7.8
△ 10.4
△ 11.8
△ 9.6
△ 9.9

△ 20.9
△ 14.8
△ 11.7
△ 15.6
△ 12.3
△ 12.5

sandy clay
loam to light
clay 2.5 Y 6/4

subsoil

weathered sandstone

grey clay

galleries and
egg chambers

△ weight per cent moisture

average from several nests
approx. 24 h after rain

Figure 3.3 Type-II mound of *Camponotus intrepidus* (after Cowan et al. 1985).

compared to the adjacent topsoil. Excavations within the mounds took the form of either chambers or galleries. Chambers were generally larger with flattened floors and vaulted roofs, whereas galleries were tubular connections between chambers and nest openings. They were concentrated within the mound and topsoil with only two or three spiral galleries extending down into the largely undisturbed clayey subsoil. The round to elliptical galleries were 13–27 mm in diameter and, within the mound, had a 2–3 mm thick dark-coloured glazed wall of finely macerated organic matter, which in the chambers increased to 5 mm in thickness. With depth, the thickness of the glaze decreased and the colour lightened, but even in the deepest galleries it could still be detected.

Chambers made by other species are known to vary greatly in size. Wheeler (1910) reported that those of the harvesting ant, *Messor barbarus* and *M. arenarius*, in North Africa were 15 cm wide and 1.5 cm high, whereas in the case of the fungi-cultivating *Atta texana* they were sometimes 50–100 cm long and 30 cm broad.

The meat ant, *Iridomyrmex purpureus*, was also active in eastern New South Wales (Cowan et al. 1985) and produced similar but somewhat larger mounds than *C. intrepidus*. The surface cover of these mounds was however very different, for it consisted of coarse particles (fine gravel size) of sandstone, ironstone and, when located near roadways, construction aggregate, glass from car windscreens, and large superphosphate pellets if the mound was near farmlands. The presence of these extraneous materials

43

indicated that at least some of the cover was derived from the surface surrounding the mound and not from the soil below the mound.

The reason for the surface cover was to provide protection against rain-splash erosion. It was shown (Mazurak & Mosher 1968) that particles larger than about 2.4 mm in diameter were rarely moved by rainsplash, and raindrops of average size were then effectively absorbed in the interstices. The mean grain size of particles on meat ant mounds was well above this figure, so that little movement was to be expected under most rainfall conditions. The same type of erosion protection has been reported in the case of bulldog ants, *Myrmecia*, from the Sydney Basin (Cowan et al. 1985) and in species of *Pogonomyrmex* and *Formica* in the western USA (Scott 1951, Mandel & Sorenson 1982).

Even though mound covers may differ from one species of ant to another they have one common characteristic: they confer longevity on the mound, and this has considerable pedological implications. Greenslade (1974) recognized a meat ant mound at least 70 years old and suggested that mound ages of 100 years were quite possible. *Camponotus intrepidus* mounds were known to last a decade and some probably persisted for several decades (Cowan et al. 1985). In this last case it was calculated that the amount of soil involved in mounding was $50 \, \mathrm{kg \, ha^{-1} \, yr^{-1}}$, or $5 \, \mathrm{g \, m^{-2} \, yr^{-1}}$. This small amount when distributed over the several decades of the mound's life becomes negligible and almost ceases to have any pedological significance. Thus, despite their prominence in the landscape, type-II mounds are much less pedologically important than the visually insignificant and ephemeral mounds of type-I.

The assessment of total bioturbation by ants is much more difficult than in the case of earthworms, for the amount of material not ejected at the surface in the form of mounds can be considerable and, furthermore, there is no ingestion of material that can be used to assess rates of mixing within the soil. This helps to explain why it is only the mound material that has been quantified in any way. Nevertheless, given this constraint, the reliability of mounding estimates differs between the two types of mounding. For type-I mounding rates, reasonable estimates are possible because the deposits can be collected on a regular and systematic basis (Humphreys 1981). Less certainty, however, surrounds type-II mounding estimates because they rely on determining rates of mound growth and decay, for which a variety of assumptions are often required. For this reason many writers have preferred to express the impact of mounding in other ways. For example Thorp (1949) observed ant activity throughout the subhumid and semi-arid areas of the USA and made a conservative estimate of the amount of soil material contained in ant mounds at any one time as being $3560 \, \mathrm{kg \, ha^{-1}}$. In Berkshire, England, Waloff & Blackith (1962) recorded the development of mounds of *Lasius flavus* over a period of several years before concluding that the total surface area could be occupied by mounds within 100 years. Baxter & Hole (1967) attempted to be more specific about the activities of *Formica cinerea montana* on a prairie soil in southwestern Wisconsin. They reported 1531 mounds per hectare, with an average diameter of 37 cm, height 15.3 cm and volume 0.02 m^3. The mounds were calculated to contain 34 m^3 ha^{-1} and to cover 1.7% of the ground surface. Taking the average occupancy time of mounds as 12 years, the whole of the surface would have been occupied by mounds in about 700 years.

Despite various shortcomings in the estimates of mounding it is nevertheless possible to gain an appreciation of the potential contribution of ants from a comparison of mounding rates (Table 3.3). At a global level the highest established rates recorded are about 10 t ha^{-1} yr^{-1} (Salem & Hole 1968, Humphreys 1981), with the higher values confined to the moister environments of the subtropics and temperate latitudes.

Termites

There are some 2000 known species of termites, which have a general tropical and subtropical distribution but do reach latitudes of 45°. They have developed the capacity to burrow and mould structures from soil and organic matter to a greater extent than any other group of soil animals. In the early literature Drummond (1884–5) suggested that termites were the tropical analogue of earthworms, but this was long before the abundance of earthworms in the tropics was appreciated.

Many termites build mounds, the density of occurrence of which varies inversely with their size; thus in Zambia small mounds can occur at the rate of 5000 ha^{-1} medium mounds 18–86 ha^{-1} and 2–12 ha^{-1} for termite hills (Pullen 1979). In Australia mound densities vary from less than 2 ha^{-1} in southern New South Wales to 210 ha^{-1} in the savanna woodland of the Northern Territory (Wood & Lee 1971). It is this feature of termite activity that is generally reported, frequently to the exclusion of everything else, even though many termites are totally fossorial. Because of the large size of many mounds, which can be up to 30 m in diameter and 10 m high, their occurrence is often taken to indicate a high rate of mounding, but these are type-II mounds, very resistant to erosion, with a considerable longevity. It has been shown (Nye 1955c, Lee & Wood 1971a,b, Pomeroy 1976a,b) that the maintenance of these large mounds required the addition of no more soil than 100–590 g m^{-2} yr^{-1}, which is less than the amount brought to the surface by many earthworms and ants. However, up to the present there has been little attempt to estimate the amount of soil termites excavate in the

Table 3.3 Rates of mounding by ants.

Location	Env. code*	Species	Rate of mounding ($t\,ha^{-1}\,yr^{-1}$)	References	Notes
Ibadan, Nigeria	TrS	Not identified	1.26–5.02	Madge (1969)	No time period for mounding specified - assumed to be one year
Berkshire, England	Tm	*Lasius flavus*	na	Waloff & Blackith (1962)	
Grant County, Wisconsin, USA	Tc	*Formica cinerea montana*	1.33	Baxter & Hole (1967)	
Southern Wisconsin, USA	Tc	*Formica exsectoides*	8.33–11.3	Salem & Hole (1968)	The higher estimate of mounding uses an average mound life of 2.2yr and is based on Andrews (1925). However, examination of Andrews indicates a 3yr period is more correct. This results in a lower rate.
Louisana, USA	Tc	*Solenopsis invicta*	1.594	Lockaby & Adams (1985)	Exotic species
Michigan, USA	Tc	*Lasius niger*	0.855	Talbot (1953)	
Mt Tatlow, USA	M	*Formica fusca*	na	Wiken et al. (1976)	
Cambridge, Massachusetts, USA	Tm	not identified	67.25	Shaler (1891)	N. B.: Bryson (1933) and Jacot (1936) state that Shaler's rate of $5.08\,mm\,yr^{-1}$ is the equivalent of $6725\,gm^{-2}\,yr^{-1}$
Petersham, Massachusetts, USA	Tm	*Formica fusca, F. neoga-gates, Aphaenogaster rubis, Lasius alienus*	0.55	Lyford (1963)	
Western Oregon, USA	Tm	*Formica sp. neoclara complex*	1.9	Forcella (1977)	Forcella (1977) assumes an average mound age of 10yr,

Table 3.3 Rates of mounding by ants.

Location	Climate*	Species	Rate	Reference	Notes
Southern Idaho, Colorado, USA	SA	Pogonomyrmex occidentalis	0.1	Sharp & Barr (1960), Thorp (1949)	Sharp & Barr show a mound growth rate of 7.2 ha⁻¹ yr⁻¹. From Thorp (1949), average mound weight is 14.4–76.2 kg and bulk density was estimated as 1.44 g cm⁻³. With these data it is possible to give an approximate value of mounding rate.
Colorado, USA	SA	Pogonomyrmex occidentalis	0.112	Mandel & Sorenson (1982)	Mandel & Sorenson do not give a mounding rate. This estimate uses the 2.8 kg average weight moved by each colony (Rogers & Lavigne 1974).
Mt Lofty, SA, Aus.	Med	Iridomyrmex purpureus	≈0.01	Greenslade (1974)	Compare to Humphreys (1985)
Cattai, Sydney, Aus.	ST	Aphaenogaster longiceps	5.45–10.69	Humphreys (1985)	On texture-contrast soil
Cordeaux, Sydney, Australia	ST	Aphaenogaster longiceps	68.38	Humphreys (1985)	
Cordeaux, Sydney, Australia	ST	Aphaenogaster longiceps	8.41–8.76	Humphreys (1981, 1985)	
Cattai & Cordeaux, Sydney, Australia	ST	Myrmecia gulosa, M. forficata, M. pyriformis, Consobrinus intrepidus	0.19–0.28	Humphreys & Mitchell (1983), Humphreys (1985)	
Sydney, Australia	ST	Consobrinus intrepidus	≈0.05	Cowan et al. (1985)	
Cattai, Sydney Basin, Australia	ST	Iridomyrmex purpureus	0.003	Humphreys (1985)	Compare to Greenslade (1974)
Killonbutta, NSW, Aus.	ST/Tc	Aphaenogaster sp.	0.28	Mitchell (1988)	
Deniliquin, NSW, Australia	SA	Chelander sps. Pheidole sp., Iridomyrmex sp. and 18 other minor species	0.35–0.42	Briese (1982)	

* See Table 3.1.

construction of subsurface galleries and move in the construction of covered runways in searching for food and moisture. The same is true of the considerable amounts of soil that termites use to pack the eaten-out portions of fallen branches, logs, standing trees and tree stumps. Thus, in the case of *Coptotermes acinaciformis*, a tree-nesting species, from southern Australia, Greaves (1962) showed that one colony had six main galleries radiating from the tree, below ground level. In section the galleries were uniformly 3 mm in height and their width varied from 6 mm to 63 mm depending on the traffic density. They were 15–28 cm below the surface and extended outwards for up to 30 m from the nest. Presumably all the excavated soil was used in above-ground packing. Ratcliffe & Greaves (1940) studied the subterranean gallery system associated with *Coptotermes lacteus* and *Nasutitermes exitiosus* in southeastern Australia. In the case of two mounds constructed by *C. lacteus* there were nine and 36 main galleries radiating from them respectively. The number of galleries associated with five mounds of *N. exitiosus* varied from 18 to 36. In general size and range, these galleries were similar to those described for *C. acinaciformis*.

With regard to non-mounded surface materials Nutting et al. (1987) estimated that for two termite species in the Sonoran Desert (*Gnathamitermes perplexus* and *Heterotermes aureus*) this amounted to $74 \, \mathrm{g \, m^{-2} \, yr^{-1}}$. In the case of covered feeding runways, Gupta et al. (1981) estimated that *Odontotermes gurdaspurensis* used $15.9–19.7 \, \mathrm{g \, m^{-2} \, day^{-1}}$ over two monitoring periods, which if applied over a six-month season would amount to $3250 \, \mathrm{g \, m^{-2} \, yr^{-1}}$, an order of magnitude greater than any other figure given for soil transport by termites. In comparison the same genus (*Odontotermes*) in an arid area in Kenya constructed similar runways at the much lower rate of $106 \, \mathrm{g \, m^{-2} \, yr^{-1}}$ (Bagine 1984). In the *Pilliga* scrub of central New South Wales (Hart 1992) surface-foraging non-mounding termites were shown to be capable of turning over the surface 10 cm of soil within a period of 260–1000 years.

In the process of excavating, transporting and depositing soil material, termites appeared to show some degree of selectivity depending upon particle size. In general, if any selectivity occurred at all, clay was preferentially accumulated in sand-rich materials and vice versa. Care had to be taken in assessing the degree to which this process had occurred, particularly in those cases where the soil consisted of an already sand-rich surface over a more clay-rich subsoil. In many instances the mound had the same particle-size distribution as that of the subsoil, that is the subsoil as a whole had been moved without any selectivity (Lee & Wood 1971a,b). However, there were certain cases in which differential accumulation of clay was undoubted. Nests of *Macrotermes natalensis* within the Kalahari sand contained 8–13% clay, whereas the clay content of the sand down to a depth of 84 m varied only in the range 1.8–5.4% (Bouillon 1970). A rather different example of clay enrichment was provided by nests of *Drepanotermes rubriceps* in South Australia, which had more clay and less sand than the surrounding surface layer of soil material. This suggested that at least some of the nest material was derived from the more clay-rich subsoil. Identification of clay minerals, however, showed that the clay of the nests was the same as that of the surface soil and differed from that of the subsoil. The termites must, therefore, have altered the ratio of clay to coarser materials (Lee & Wood 1971a,b).

Stoops (1964) showed that particle sorting by termites may become rather more complex. Mounds of *Cubitermes* spp., compared with neighbouring soils, were enriched in particles of <100 µm and >500 µm, with a considerable impoverishment of particles of intermediate size. Stoops showed that the termites swallowed fine soil particles, carried them in the crop and then regurgitated them, whereas larger particles were carried in the mandibles. It was concluded that particles in the 100–500 µm range were too large to be swallowed and too small to be conveniently carried.

In the process of moving soil material around and mixing it with organic matter in various forms in order to build nests, runways and galleries, termites are responsible for the development of new fabrics. It is possible to differentiate between those fabrics that are mostly mineral and those dominantly organic. The mineral fabrics make up the greater part of the outer walls of mounds and the infilling of former galleries, whereas the organic-dominated fabrics occur in nursery areas and as gallery linings (Lee & Wood 1971a,b).

Despite all its possible shortcomings Table 3.4 lists data on termite bioturbation in terms of rate of mounding. It shows that termite activity is most pronounced in drier environments, especially in semi-arid to wet/dry tropics. The common upper limit of mounding is $1\,t\,ha^{-1}\,yr^{-1}$ and in only two cases have much higher rates been determined. However, these figures do not include the amount involved in the excavation of subsurface galleries, the sheeting of surface runways and surface packing, which, as mentioned previously, can involve much higher rates of soil transport than that involved in constructing termitaria.

Other invertebrates
There are many other invertebrates that spend at least part of their life cycle in the soil, alter its fabric and may contribute to mounding. Hole (1981) listed beetles, bees and wasps, cicadas, crickets, grasshoppers, centipedes, spiders, scorpions, woodlice and crayfish. Although a great deal is known about some pest species, such as the Australian plague locust (Key 1959), there are almost no studies where the effect on the soil is of primary interest.

Table 3.4 Rates of mounding by termites.

Location	Env. code*	Species	Rate of mounding ($t\,ha^{-1}\,yr^{-1}$)	References	Notes
Ibadan, Nigeria	TrS	*Macrotermes nigeriensis*	1.260	Nye (1955c)	The rate normally quoted from Nye is the rate of subsoil mounded, which is about one-third the rate given
Senegal	TrS	*Macrotermes subhyalinus*	0.800	Pomeroy (1976a,b)	
Zaria, Nigeria	TrS	*Trinervitermes oeconomus, T. ebenenanus*	na	Leow & Degge (1981)	
Kampala, Uganda	TrS	*Macrotermes bellicosus, Cubitermes ugandensis, Pseudacanthotermes* sp.	na	Pomeroy (1976b)	
Comoe, Ivory Coast	TrS	*Macrotermes bellicosus*	na	Lepage (1984)	Assumes mound growth rate of $0.3\,m^3\,yr^{-1}$
Marsabit, Kenya	SA	*Odontotermes* sp.	1.059	Bagine (1984)	Rate of soil sheeting
Lumbumbashi, Zaire	TrS	*Cubitermes* sp.	3.2–5.9	Aloni & Soyer (1987)	
Tucson, Arizona, USA	SA	*Heterotermes aureus, Gnathamitermes perplexus*	0.744	Nutting et al. (1987)	Rate of packing above-ground; field experiment
San Carlos de Rio Negro, Venezuela	TrH	7 to 14 species depending on soil type	0.01–0.78	Salick et al. (1983)	Rate lowest on podzols and highest on laterite
Kurukshetra, India	TrS	*Odontotermes gurdaspurensis*	32.500	Gupta et al. (1981)	Rate of soil sheeting. Assumes rate of $17.8\,g\,m^{-2}\,day^{-1}$ applied for 6 months
Brocks Creek, NT, Australia	TrS	*Tumultitermes hastilis, T. pastinator, Nasutitermes triodiae*	1.250	Williams (1968)	

Table 3.4 Rates of mounding by termites.

Location	Climate*	Species	Rate	Reference
Adelaide R., NT, Australia	TrS	*Nasutitermes triodiae*	1.150	Lee & Wood (1971b, Site 4)
Gerry Airfield, NT, Australia	TrS	*Nasutitermes triodiae, Tumultitermes hastilis, Amitermes vitiosus, Drepanotermes rubiceps*	4.700	Lee & Wood (1971b, Site 6)
Charters Towers, Queensland, Australia	SA	*Nasutiterms longipennis, Tumultitermes pastinator, Amitermes vitiosus, Drepanotermes rubiceps*	0.4–0.8	Holt et al. (1980)
Englebrookk, SA, Australia	Med	*Nastutitermes exitiosus*	0.020	Lee & Wood (1968)
Cattai, NSW, Australia	ST	*Nastutitermes exitiosus*	na	Humphreys (1985)

* See Table 3.1.

However, some work has been undertaken: Thorp (1949) described crayfish chimneys and burrows in Indiana, Williams (1974) dealt with crayfish in a Tasmanian moorland, and Greenway (1980) recorded burrows of the Australian arid zone crab to a depth of over 50 cm. A detailed investigation of soil disturbance by isopods in the Negev Desert of Israel (Yair & Rutin 1981) showed them to be the source of much of the material subsequently removed as sediment from a small catchment.

More is probably now known about the cicada where both subsurface burrowing and surface mounding have been observed (Humphreys 1985, 1989). Figure 3.4 is an example of a cicada burrow with a very heterogenous infill, which includes litter and rainwash as well as ant and termite deposits. It was constructed by a cicada nymph as it moved towards the surface. When the organic-rich surface was reached, that material was used to plaster the burrow wall. At the same time pellets 5–10 mm long and 3–6 mm in diameter were formed, which were possibly used after the cicada reached the surface to construct a turret, which can be up to 95 mm high, with an external basal diameter of 60 mm tapering to 45 mm and weighing 120 g when air-dried. Such turrets are rarely found, for they are type-I mound material and are extremely fragile. This could lead to a lack of their recognition and a gross underestimate of the amount of material deposited in this way – a situation that can only be exacerbated by the periodic and in many ways unknown nature of the cicada's life cycle.

Table 3.5 gives data on the rate of mounding by a range of these lesser known invertebrates. Rates are generally less than $0.5\,t\,ha^{-1}\,yr^{-1}$, although Kalisz & Stone (1984) reported mounding of up to $1.85\,t\,ha^{-1}\,yr^{-1}$ for scarabs and two studies on crayfish reported substantially higher rates of between 6.3 and $8.4\,t\,ha^{-1}\,yr^{-1}$ in very localized and mostly waterlogged sites (Thorp 1949, Young 1983).

These results, however, must be considered somewhat artificial for all the species examined have been considered as if they were operating in isolation whereas in almost every case investigated several animals were operating together to cause bioturbation. Humphreys (1985) has attempted to assess the overall rate of bioturbation, both surface and subsurface, at two sites (Cattai & Cordeaux, Fig. 3.1) in the Sydney region of New South Wales. The results for surface mounding are summarized in Table 3.6. This represents a compilation that varies not only in the period and type of measurement, but in quality too. However, this remains as one of the few sites from which data of this type are available.

It is clear that the dominant mounder is *Aphaenogaster longiceps*. In contrast mounding at both sites by other ant species, including type-II mounders, remains low. The rate of mounding by termites is very low, with

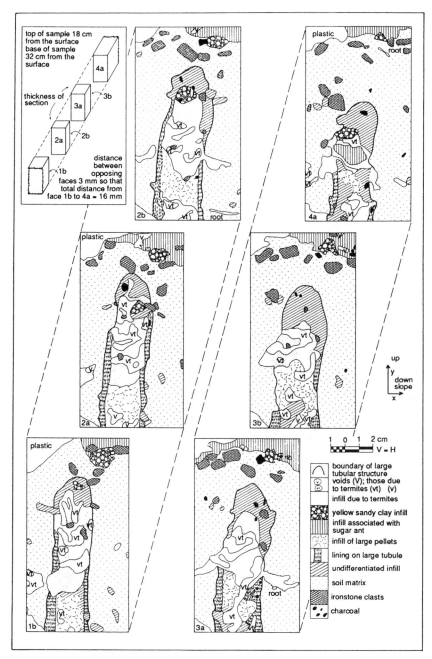

Figure 3.4 Cicada burrow infill (after Humphreys 1985).

Table 3.5 Rates of mounding by other invertebrates.

Location	Env. code*	Species	Rate of mounding ($t\,ha^{-1}\,y^{-1}$)	References	Notes
Southern Indiana, USA	ST	Crayfish (Cambarus spp.)	6.3–8.4	Thorp (1949)	
Kansas, USA	Tc	Various insects including ants, wasps, ground beetles, etc.	2.360	Bryson (1933)	
Kansas, USA	Tc	Dung beetles (Copris tullius, Pinotus carolinus)	0.160	Lindquist (1933)	Jacot (1936) gives a rate of dung burial as $2.35\,g\,m^{-2}\,yr^{-1}$
Florida, USA	ST	Scarab beetle (Peltrotupes youngi)	0.01–1.85	Kalisz & Stone (1984)	
Negev, Israel	A	Woodlice, isopod (Hemilepistus reaumuri)	0.0–0.41	Yair & Rutin (1981)	Activity increases with rainfall in the range of 65 to 310 mm yearly
Negev, Israel	A	Desert snail (Sphincterochila zonata)	0.0–0.06	Shachak & Steinberger (1980)	Soil turnover is confined to the upper few millimetres where an algal mat is ingested
Cattai & Cordeaux, NSW, Australia	ST	Cicadas (Psaltoda moerens, Thopa saccata)	0.03–0.19	Humphreys & Mitchell (1983), Humphreys (1985)	
Woronora, NSW, Australia	ST	Crayfish (Eustacus hierensis)	7.300	Young (1983)	
Cattai, NSW, Australia	ST	Trapdoor spider (Arbanitis sp., Dyarcyops sp.)	0.00–0.03	Humphreys (1985)	

* See Table 3.1.

Table 3.6 Mounding rates by mesofauna at Cattai and Cordeaux.

Fauna type	Species	Mounding type (see text)	Rate of mounding (t ha^{-1} yr^{-1}) Cattai	Cordeaux
Funnel ant	*Aphaenogaster longiceps*	I	5.45	8.41
Ant	*Camponotus intrepidus*	II	–	0. 19
Bulldog ant	*Myrmecia forficata*			
Bulldog ant	*M. pyriformis*			
Ant	*Camponotus intrepidus*	II	0.28	–
Bulldog ant	*Myrmecia gulosa*			
Termite	*Heterotermes ferox*			
Cicada	*Psaltoda moerens*	I	0. 19	
Cicada	*Thopha saccatta*	I	–	0.03
Trapdoor spider	*Arbanitis* sp.			
Trapdoor spider	*Dyarcyops* sp.	I	0.03	–
Earthworm	*Cryptodrilus* sp.			
Earthworm	*Oreoscolex* sp.	I	–	1.33
	Total		5.95	9.96

Heterotermes ferox constructing an inconspicuous mound, although in very similar adjacent areas, *Nasutitermes exitiosus* and *Coptotermes lacteus* construct sizable mounds (Ratcliffe et al. 1952). Turrets are sometimes constructed by cicada nymphs (Humphreys 1989); however, an estimate of the amount of material involved is no more than a very rough estimate. More certainty exists with the soil excavated by trapdoor spiders, but their burrows are much less common than the cicada burrows of similar size. Earthworm activity was recorded at Cordeaux, where casts were easily collected (Humphreys 1981). At Cattai earthworms also occurred but the dense leaf litter and ground cover prevented systematic collection.

Detailed fabric analysis of texture-contrast soils, at Cattai, down to a depth of 1 m, provided knowledge of subsurface bioturbation (Humphreys 1985, 1994a), which when added to that of surface mounding gives some idea of the overall pattern of bioturbation. From the evidence provided by tubules and relic tubules (Fig. 3.5) it is apparent that bioturbation is mostly confined to the topsoil (95%), which is confirmed by the concentration of charcoal and stones at the base of the topsoil. Bioturbation was shown to extend into the subsoil, but only to the extent of 5% of the total, and by far the greater part of this was concentrated within 5 cm of the topsoil/subsoil boundary. Only half of the bioturbated subsoil was transported into the topsoil and the greater part of this ended up as a component of the surface mounds.

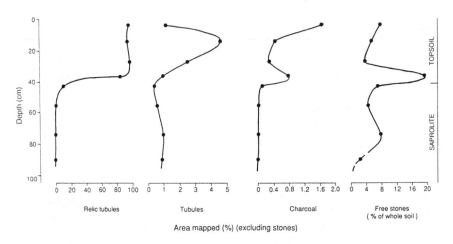

Figure 3.5 Fabrics in texture-contrast soils (after Humphreys 1985).

Vertebrates

Birds
Data on the role of vertebrates in soil bioturbation are much more sparse and patchy than in the case of the invertebrates. This can be illustrated by the situation in Australia with regard to the possible disturbance of soil by birds. Of the 700 species of Australian birds (Frith 1976), at least ten extensively probe the soil during feeding, 11 rake the forest litter, 26 nest in soil burrows, 44 nest in shallow scrapes on the ground often in colonies, three build large incubating mounds, and ten construct mud nests in trees. Not all of these are likely to be important as soil bioturbators, but some occur in such large numbers that they could be very significant. However, even in the case of seabirds that congregate in such large numbers on islands and where soil disturbance as a result of nest-making is inevitable, the only data available have to do with nutrient cycling and the accumulation of guano. In only one case have sufficient data been reported for a reasonable evaluation of soil disturbance to be made and this is for the lyrebird (*Menura novaehollandiae*; Adamson et al. 1983, Mitchell 1985).

Lyrebirds are moderately large ground feeders, found in a range of vegetation types in the more humid areas of southeastern Australia, up to an elevation of 1500 m. Soil is disturbed in the process of feeding and also when the male bird constructs display mounds. These activities were investigated at Blackheath in the Blue Mountains of New South Wales (Fig. 3.1). The site had an elevation of 900–1000 m, with an open eucalypt forest and a moderately dense understorey (Fig. 3.6). The display mounds were con-

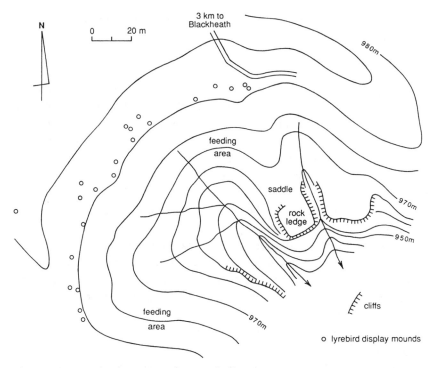

Figure 3.6 Lyrebird activity (after Mitchell 1985).

structed within 15–20 m of the plateau edge, with feeding areas confined to the slopes and benches immediately below. The mounds, 22 in number, were about 20 cm high, circular in plan and each of them covered an area of 1 m². They were fringed by a band 10–20 cm wide where the soil was disturbed to a depth of 10 cm. The total amount of material involved in any one mound was only some 25 kg and, as no more than two of them were replaced in any one year, the amount of soil detachment resulting from mound construction on an annual basis was negligible.

The amount of material detached as a result of lyrebird feeding was much greater and was concentrated wherever moisture conditions encouraged a build-up in the invertebrate population, as on the slopes below the display mound area. The birds removed much of the vegetation in the process of digging through the coarse dry surface litter into the moister comminuted litter and mineral soil. On steeper slopes this created a vertical face 5–10 cm high, which was advanced by repeated working, to form a zone of excavation, with a debris plume, up to 1.5 m long, thrown out behind, which caused the slope as a whole to have a terraced appearance. On gentler slopes where direct downslope movement of debris did not occur, feeding patches were more circular and covered several square metres. Mitchell (1985)

57

calculated that as a result of these feeding activities the total rate of disturbance was $6300\,\mathrm{g\,m^{-2}\,yr^{-1}}$, of which $4500\,\mathrm{g}$ was mineral soil.

Mammals

The same paucity of data with regard to soil disturbance is to be seen in the
case of mammals. However, there can be little doubt that mammals must
contribute significantly to such disturbance, for out of a total of 777 genera
of terrestrial mammals (Walker 1968), 447 or 58% of them could be
regarded as potentially significant in soil disturbance.

It is possible to divide this group of 447 genera into three subgroups,
which:

1. disturb soil by surface feeding (141 or 31% of potential bioturbators)
2. disturb soil by surface feeding and burrowing (58%)
3. disturb soil because of their subterranean lifestyle (11%).

These three groups show a decrease in body size from the large herbiv-
ores, through badgers and wombats, to the fossorial moles. Almost all the
data available on these mammals are of an ecological nature, so that the
amount of associated soils information increases from groups 1 to 3 as the
mammals involved become smaller and more and more closely associated
with the soil.

In the case of group 1 the outstanding example (excluding humans,
domestic stock and feral animals) is provided by the elephant. Giardino
(1974) observed a great increase in the amount of surface runoff and gully-
ing within the Luangwe National Park, Zambia, paralleling an increase in
the elephant population, which led to a stripping of the vegetation and an
increased trampling of the soil.

In the case of group 2 mammals their contact with the soil is more inti-
mate for they excavate burrows as homes and hence their potential for soil
disturbance is much increased. Typical examples are provided by the badgers of North America and Europe (Neal 1948) and aardvark in Africa
(Melton 1976). More pedological data are available for the common wombat (*Vombatus ursinus*) in Australia, which operates in much the same way,
combining burrow construction and surface feeding. McIlroy et al. (1981)
showed that burrows had a mean length of 10 m and the amount excavated
averaged $1.2\,\mathrm{m^3}$. In all 32 ha was examined in this study, within which the
total burrow volume was $18\,\mathrm{m^3}$. This was equivalent to $84\,\mathrm{g\,m^{-2}}$ of soil, the
pedological significance of which becomes much less when spread over the
several decades during which the burrow was in use. This conclusion was
supported by subsequent work in the Snowy Mountains (Mitchell 1985)
where turnover from wombat burrowing was assessed at $10\,\mathrm{g\,m^{-2}\,yr^{-1}}$ at a
maximum. It was, however, during surface feeding that most soil was

disturbed, for the wombats grazed partly on plant roots and in the process excavated soil to a depth of 8–15 cm. This activity appeared to be random, but once initiated, patches up to 250 m^2 were worked, over several years. Detachment rates varied from 183 to 463 g m^{-2} yr^{-1}, which is 20–50 times the burrowing rate.

The problem of assessment of soil disturbance is rather different for some of the smaller mammals of group 2 such as ground squirrels, chipmunks, marmots and prairie dogs, which are very widely distributed throughout North America and Eurasia from the tundra to the arid zone (Khodashova & Dinesman 1961). These small mammals are thought by many to be major contributors to soil disturbance, not only because of their wide distribution, but also because of the scale of their burrow construction. Thus, the average-size prairie dog "town" varies from 13 ha to 23 ha (Bishop & Culbertson 1976), and Sheets et al. (1971) described a "town" of 43 ha in South Dakota, containing 575 animals and 139 active burrows, which varied in length from 13 m to 33 m, in diameter from 10 to 12.5 cm, and extended to a depth of 2 m. However, this is the only way these mammals disturb the soil, for they are herbivorous rodents and are primarily surface collectors of food. So that, in determining soil turnover rate in this case, burrowing is the only thing that has to be taken into account and when the longevity of the burrows is considered, the pedological significance of the soil turnover rate becomes rather less.

The smaller fossorial mammals of group 3 that spend their entire life underground are in a very different situation, for they are concerned with soil disturbance throughout the whole of their active lives. This group of small mammals includes both the carnivorous moles and shrews as well as herbivorous gophers, mole rats and sand rats. They occur throughout Europe (Godfrey & Crowcroft 1960, Mellanby 1971) and spread into the Russian steppes (Zlotin & Khodashova 1980) and Anatolia (de Meester 1970). They also occur throughout North America (Price 1971) and analogous species have been reported from South America (Reig 1970) and Africa (Genelly 1965, Gakahu & Cox 1984).

De Meester (1970) showed that the feeding burrows of these fossorial mammals were very extensive with single burrows extending over 1.5 ha and reaching a depth of 4 m. Burrows were often plugged either to hinder predators or to dispose of waste products, or simply as a means of dumping excavated soils. The back-filling with soil, which formed tubes of material of a different character from their surroundings, **krotovinas**, was at times so common as to almost obliterate any sign of the original soil fabric. The back-filling activities of these fossorial mammals also frequently resulted in the soil being deposited as mounds, which covered from 10 to 25% of the

Table 3.7 Rates of mounding by vertebrates.

Location	Env. code*	Species	Rate of mounding ($t\,ha^{-1}\,yr^{-1}$)	References	Notes
Moscow, USSR	Tc	Mole (Talpa europaea)	3.9–18.6	Abaturov (1972)	
Caspian Region, USSR	SA	Gopher (Citellas pygmaeus)	1.5	Abaturov (1972)	
Karpacz, Poland	Tc	Mole (Talpa europaea)	1–3	Jonca (1972)	
Ave et Auffe, Belgium	Tm	Badger (Mele meles) Rabbit (Oryctolagus cuniculus)	0.009	Voslamber & Veen (1985)	
De Blink, Netherlands	Tm	Rabbit (Oryctolagus cuniculus)	0.81	Rutin (1992)	Rate estimated from data in paper
Ardennes, Luxembourg	Tm	Mole (Talpa europaea) Vole (not identified)	1.94	Imeson (1976)	
Florida, USA	ST	Pocket gopher (Geomys pinetus)	0.29–8.16	Kalisz & Stone (1984)	At these sites scarab beetles mound at rates of 1.5–3.7 $t\,ha^{-1}\,yr^{-1}$
Texas, USA	SA	Pocket gopher (Geomys breviceps brazensis)	0.81–15.87	Buechner (1942)	
California, USA	M	Pocket gopher (Thomomys monticola)	0.02	Grinnell (1923)	nb. Values averaged over Yosemite National park. Locally much higher rates occur.
California, USA	M	Pocket gopher (Thomomys monticola)	16.8	Ingles (1952)	
San Francisco, USA	Med	Pocket gopher (Thomomys bottae)	na	Black & Montgomery (1991)	
Utah, USA	M	Pocket gopher (Thomomys talpoides moorei)	11.2–14.6	Ellison (1946)	

Table 3.7 Rates of mounding by vertebrates.

Location	*	Animal	Rate	Reference	Notes
Niwot Ridge, Colorado, USA	M	Pocket gopher (*Thomomys talpoides*)	3.9–5.8	Thorn (1978)	
Wollowa, USA	M	Various mammals, probably pocket gopher	0.25	Stine (1974, quoted in Yair & Rutin 1981)	
Olympus National Park, USA	M	Various mammals, probably pocket gopher	0.48	Stine (1974, quoted in Yair & Rutin 1981)	
Yukon, Canada	P	Arctic ground squirrel (*Citelius undalatus*)	0.34	Price (1971)	nb. Values average over the whole site. Local squirrel activity produces up to 18 t ha^{-1} yr^{-1}
Rajasthan Desert, India	A	Desert gerbil (*Meriones hurrianae*)	1.04	Sharma & Joshi (1975)	
Negev, Israel	A	Porcupine (*Hystrix indica*)	0.029–0.302	Yair & Rutin (1981)	
Killonbutta, NSW, Australia	ST/Tc	Echidna (*Tachyglossus aculeatua*)	1.65	Humphreys & Mitchell (1983), Mitchell (1988)	
Blackheath, NSW, Australia	Tm	Lyrebird (*Menura superba*)	0.4	Humphreys & Mitchell (1983), Mitchell (1988)	Locally, mounding rate to 44.7 t ha^{-1} yr^{-1}

* See Table 3.1.

total surface. Imeson & Kwaad (1976), in a forested area of the Ardennes, showed that moles and voles provided $2000 \, g \, m^{-2} \, yr^{-1}$ to the surface, which was more than sufficient to provide all the material being transported out of the area by fluvial processes. Zlotin & Khodashova (1980) estimated that the sand rat of the Russian steppes, as a result of extensive burrowing and mounding, could turn over the surface 20 cm of soil in 2000–5000 years.

The small body size of the fossorial mammals introduced the possibility of sorting, particularly where gravel fragments occur. Johnson (1989) showed that, in certain gravelly soils in California, pocket gophers could not move gravel larger than 7 cm in diameter, so that such gravel fragments settled to the lower limit of gopher activity, leaving behind a surface layer containing gravel fragments less than 7 cm in diameter only. Thus, a two-layer soil was produced as a result of the burrowing activities of these small mammals.

Practically all of the available data on bioturbation by these small fossorial mammals are in terms of the rate of mounding, so that it is at best a minimal estimate of the total soil material disturbed (Table 3.7). Generally mounding rates range from 1 to $15 \, t \, ha^{-1} \, yr^{-1}$, although higher rates of up to $20 \, t \, ha^{-1} \, yr^{-1}$ have been recorded (Ingles 1952, Price 1971, Abaturov 1972).

Plants

Soil bioturbation by plants – or floralturbation according to Schaetzl et al. (1989) – is most evident as a result of tree fall (or uprooting), a fact that has been emphasized since the time of Shaler (1891) and in many subsequent studies (Lutz & Griswold 1939, Schaetzl et al. 1989, 1990). From our own experience (Humphreys & Mitchell 1983) it is possible to recognize two distinct forms of soil disturbance resulting from uprooting: one in which a horizontally extensive slab of material is detached, and another where the detached mass forms a root-ball. The significance of this distinction can be understood with reference to the following examples. A typical soil slab detached by the fall of a tree with a 16 m trunk, an 8 m diameter crown and a girth of nearly 2 m, measured $4 \, m \times 2.6 \, m$ and was up to 0.15 m thick. In addition broken lateral roots ruptured the surface for 1.5 m beyond the pit created by the slab and as a result intact masses of soil weighing 10–17 kg were thrown almost 3 m down slope. In this fall 2.5 t of material was disturbed and a pit and mound microtopography created. Root-ball detachment from a smaller tree (1.2 m girth; Fig. 3.7) involved a near-hemispherical body of material, about 1.4 m in diameter and weighing 0.4 t. Microtopography was not so marked as in slab detachment, for in this case the cone merely rotated in place (Mitchell 1985). In both cases the soil material

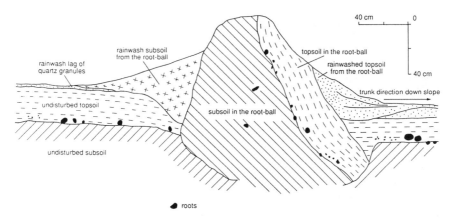

Figure 3.7   Root-ball failure (after Mitchell 1985).

detached by treefall was rather more resistant to subsequent downslope lateral movement than the products of animal bioturbation because of the persistence of plant roots (Terwilliger & Waldron 1990). A natural consequence of this is that downslope movement of material when detached by treefall is restricted. Thus, Denny & Goodlett (1956), from a site in Pennsylvania, estimated that up to 112 m³ of material was detached over an area of 270 m² during a forest regeneration cycle of 200 years, but that only half of this was moved as far as 1 m down slope, as a result of initial disturbance. Even though this was probably an overestimate (Schaetzl & Follmer 1990) it still suggested that even under optimum conditions little soil was displaced more than 4 m. A similar conclusion was reached by Burns & Tonkin (1987) in a New Zealand subalpine *Nothofagus* forest, where pit and mound microtopography covered 15–18% of the area and yet the actual displacement of material was restricted to the equivalent of a layer only 11–13 mm thick, which moved less than 2 m down slope.

However, the relationship between treefall and soil detachment is not very simple or direct, for dead treefalls cause little if any soil disturbance compared to the fall of a tree with a living root system. This is illustrated by the incidence of treefall and associated soil detachment at nine sites in New South Wales (Fig. 3.1), which were observed over a two to three year period (Table 3.8), where treefall occurred at every site and yet significant soil detachment was confined to three sites only (Lidsdale, Welumba Creek and Link Trail). Killonbutta provides the most extreme example for, despite a very high rate of 210 treefalls, soil detachment was hardly measurable, for they were all dead falls.

In addition to soil detachment as a result of treefall all plants, but especially trees, are capable of exerting pressure on the surrounding soil by way

63

Table 3.8 Treefall and soil detachment (after Mitchell 1985).

	Fallen trees	% responsible for soil detachment	Rate of soil detachment $(gm^{-2}yr^{-1})$
Cattai	7	–	7
Cordeaux	7	–	19
Blackheath	16	14	17
Lidsdale	145	68	104
Killonbutta	210	1	0.4
Snowy Mountains			
Welumba Creek	60	83	77
Link Trail	29	54	45
Clover Flat	61	20	18
Wolsley's Gap	56	11	19

Table 3.9 Root biomass (tha^{-1}) in various forests.*

A.	Tropical lowland forests	
	South America	9–132
	Central America	2–16
	Africa	16–52
B.	Tropical montane	
	Central America	7–72
	Africa	9+
	Far East	40
C.	Drier tropics	
	Central America	5–45
	Africa	6–12
D.	Temperate forests (Eurasia–North America)	
	Douglas fir	14–88
	Spruce	25
	Pine	24–44
	Fir	16–44
	Oak	18–74
	Beech	51
E.	Mangrove	
	Underground roots	50–190
	Prop roots	3–116
	Total roots	147–510

* Modified from: Bancalari & Perry (1987), Castellanos et al. (1991), Cavelier (1992) and Komiyama et al. (1987).

of root wedging and stem growth (van Hise 1904). Tree roots comprise 9–50% of the plant's biomass (Retzer 1963, Odum 1971) and, as they grow, they displace soil material, commonly at rates $>500\,g\,m^{-2}\,y^{-1}$ (Coleman 1976), a process in which even root hairs are involved (Blevins et al. 1970). The results of growth pressures are most clearly seen where trees develop massive trunks and the consequent radial pressures exerted on the soil lead to the development of butt hillocks around the base of the stem (Butuzova 1962, Ryan & McGarity 1983). Because of the concentration of rainfall as stemflow and its ease of entry into the surrounding butt hillocks, it follows that the occurrence of such hillocks can have a profound effect on the flow of water across particular landscapes.

Nevertheless, the relative importance of this means of soil displacement cannot be other than very small for the total below-ground biomass in many different forest types rarely exceeds $100\,t\,ha^{-1}$ (Table 3.9), which amounts to <1% of the volume of the upper 1 m of soil.

Conclusion

On a global scale it can be concluded that by far the greater part of soil bioturbation is the result of animal activity. In general it appears that plants play much more of a protective role towards the surface of the lithosphere (see Ch. 4) rather than aiding in its breakdown, even though treefall-induced pit and mound topography may persist for decades or even a few centuries.

From these data, global rates of faunalturbation can be compiled (Fig. 3.8). These show that earthworms are the most important bioturbators, followed by the ants and small fossorial mammals, with the effect of termites and other invertebrates being somewhat less. Although all the data on which these conclusions are based are both partial and very approximate and a great deal more data are waiting to be collected, all that will possibly happen in the future is that the amount of perceived bioturbation will increase, but the relative importance of the groups of fauna are unlikely to change.

The impact of these various major groups of bioturbators is not uniform across the Earth's surface because they each have their own environmental preferences (Fig. 3.9). Termites are for the most part tropical and subtropical although they do extend to 45° north and south (Lee & Wood 1971a). In contrast earthworms extend to the permafrost but prefer moist environments of grasslands and forests (Edwards & Lofty 1977, Lee 1985). Ants are more widespread than termites and earthworms, ranging from deserts to forests and into high mountains (Wheeler 1910, Pisarski 1978). The small fossorial mammals are also widespread, extending into the tundra and the alpine zone not favoured by invertebrates. In addition there are differences within the same group depending on the major environment in which they are operating (Table 3.10, Fig. 3.9). It is also to be noted that bioturbation can alter the sedimentological character of the soil where it has the potential to be sorted. It needs to be stressed how tentative and approximate are all

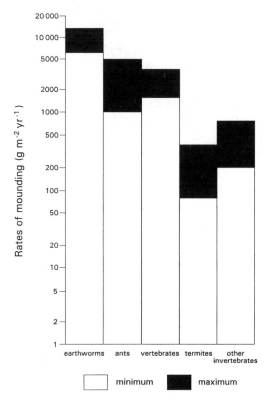

Figure 3.8 Generalized rates of mounding.

these conclusions and that anything to do with rates is probably a gross understatement. Nevertheless, there is every reason to expect that at any natural site the combined effect of bioturbation will be considerable, in terms of both the rate at which soil is mixed below the surface and deposited (i.e. mounds) on the surface, and the formation of pedotubules, krotovinas

Table 3.10 Relative status of different fauna groups in different environments (*italics* indicate where more data are available).

Env. code*	Fauna groups
P/M	*vertebrates* > ?
Tm	*earthworm* > *ants* > *vertebrates*
Tc	*earthworms* > *vertebrates* ≥ *ants* > other invertebrates
Med	earthworms > vertebrates > termites ≥ ants
SA	vertebrates > termites ≥ ants
ST	ants = earthworms = vertebrates > termites
TrS	*earthworms* > *termites* = *ants*
TrH	*earthworms* > termites
A	vertebrates > other invertebrates

* See Table 3.1.

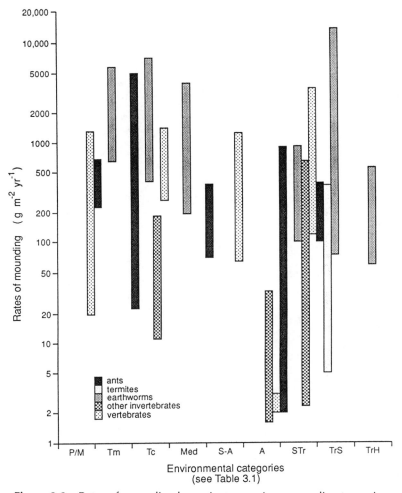

Figure 3.9 Rates of mounding by various organisms according to environment.

and other fabrics. In brief, much of the near-surface soil, the bioturbated mantle, or biomantle (Johnson 1990), owes its character to these soil-moving organisms. Furthermore, the mounds represent temporary storage sites of sediment that can be acted upon by other processes such as rain-splash erosion. This theme is developed in the next chapter.

CHAPTER 4

Rainwash

Consistency in opinion is the slow poison of intellectual life, the destroyer of its vividness and energy. *Humphrey Davy*

In Chapter 3 the formation of a soil mantle as a result of bioturbation was discussed: the biomantle of Johnson (1990). Consideration will now be given to the effects on this mantle of rainsplash, slopewash and sediment rafting, which can be collectively referred to as **rainwash**, a process responsible for detachment, transport, sorting and redeposition of near-surface materials.

The degree to which rainwash can operate is generally severely curtailed by the presence of vegetation cover. However, there are situations where special conditions have intervened to remove vegetation, or maintain vegetation-free zones, and they will be considered first of all. Thorp (1949) described such an effect on the Great Plains of the USA, where 20–50 ant mounds per hectare occurred, each of which was surrounded by a circular zone of bare soil up to 10 m wide (16–40% bare ground). Wheeler (1910) observed clearing well beyond such mounds in the form of paths 10–20 cm wide and 20–30 m long. On the Loita Plains of southwestern Kenya (Glover et al. 1964) termite mounds up to 60 cm high occurred at a density of some 650 km^{-2}. They varied from being circular, 6.5–10 m in diameter, on the interfluves, to ellipsoidal, and up to 14 m long, on the slopes down to the drainage lines, which gave rise to a peacock feather pattern visible on aerial photographs (Fig. 4.1a). The downslope-pointing tails of the ellipsoids were practically bare of vegetation (Fig. 4.1b) because the material from the termite mounds was easily dispersed and had formed a hard compact surface layer on being washed down slope. With such a concentration of mounds the effect on runoff characteristics must be considerable.

In western Queensland a succession of favourable seasons was responsible for a prolific growth of native grasses, which in turn caused the termite (*Drepanotermes perniger*) population to increase (Watson & Gay 1970). In ensuing seasons of poor grass growth the swollen termite population, in harvesting the sparse grass cover, exposed the surface to more effective processes of erosion. Erosion in turn exposed the hard capping of the

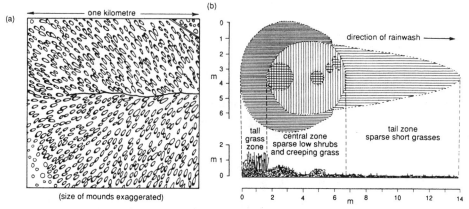

Figure 4.1 Termitaria, Kenya (after Glover et al. 1964): (a) general distribution;
(b) single termite mound.

termite nests, shaped like an inverted saucer some 2–3 m in diameter, lying immediately below the shallow, friable topsoil and covering about 20% of the ground surface. Being rock-hard and persistent they formed centres for runoff initiation.

Over great areas of the Earth's surface, especially in periodically dry environments, the most widespread and efficient manner of removing protective vegetation and producing bare surfaces susceptible to erosion is by means of fire. The burned surfaces may also develop water repellency (de Bano et al. 1970, Savage 1974, Reeder & Jurgenson 1979, Mitchell & Humphreys 1987), which is of additional pedological significance for it restricts infiltration of even low-intensity rains into materials that have a field texture grade as light as loamy sands. The erosional processes that occurred on these bare surfaces after bushfires were investigated in detail within the Sydney Basin by Mitchell & Humphreys (1987). Figure 4.2a is an idealized section of part of a gentle hillslope surface immediately after a bushfire, from which a considerable amount of the ash was removed by the wind, before the surface was subject to particle detachment by raindrop impact. Evidence of such detachment was seen in mineral particles adhering to bark and leaves of plants for up to 50 cm above the ground, and in the formation of soil pillars up to 5 cm high capped by gravel, twigs and leaves.

Subsequent overland flow occurred in very shallow, ephemeral, anastomosing channels, not more than a few centimetres wide, which was made more turbulent by raindrop impact (Emmett 1970). Despite the shallowness of such flows, generally not more than a few millimetres deep, it was possible to distinguish the same kind of materials in transport as in the much more substantial channel flows of rivers and streams (Moss & Walker

70

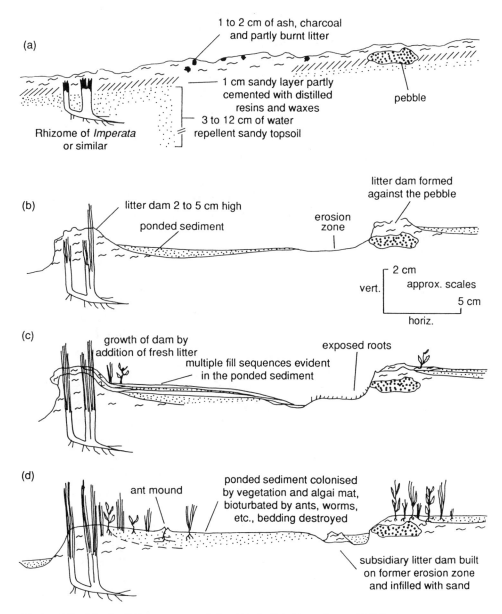

Figure 4.2 The bushfire sequence (after Mitchell & Humphreys 1987):
(a) immediately post-fire; (b) after the first runoff; (c) after several runoff events;
(d) litter decay stage.

1978). First the organic debris, leaves, twigs and bark, together with copious amounts of charcoal, were transported as **floating load**. Secondly the coarser, sand-size, mainly quartz particles were moved by rolling and saltation as **bedload**. A minor amount was also transported as **suspended load**, which was indicated by turbidity in the flow in the first few minutes of any runoff episode (Walker et al. 1978, Moss & Walker 1978, Mitchell & Humphreys 1987). Sampling of the turbid flow showed that up to 80% of the suspended material consisted of silt-and clay-size particles, which because of their fine grain size moved farther and faster than the floating and bedloads, which remained behind on the hillslope.

Normally the fines are removed completely from the hillslope system, but there are small isolated areas where they are retained, so that the process of lateral sorting is readily discernible over quite short distances. Mitchell & Humphreys (1987) described such a situation across a 20 m wide, gently sloping sandstone bench, where the topsoil was underlain across a sharp boundary by a saprolitic sandy clay. The topsoil varied from a 150 mm thick layer of single-grain medium sand, where the sand gained access to the bench, to a 25 mm thick layer of clayey fine sand, near the outer margin of the bench (Fig. 4.3b,c). Even without fire, litter moved downhill as an independent layer both as floating load and also more slowly and continuously as a result of the expansion and contraction of individual components on wetting and drying (Imeson & van Zon 1980, Mitchell 1985). Nevertheless, after fires, litter was the major component of material moving down slope in response to overland flow (Blong et al. 1982, Atkinson 1984). Such litter contained a great number of elongate fragments of eucalypt leaves, needle-shaped leaves of *Casuarina* and *Hakea* and strips of bark and twigs, which lodged between obstructions caused by rocks, plant stems and partly burnt tussock grasses to form a dam usually 2–5 cm high, with a downslope convex plan form and a spacing between the dams that varies from 0.2 m to 2 m (Mitchell & Humphreys 1987; Fig. 4.4; Plate 2). There was a tendency for these litter dams to coalesce and form a feature several metres long, running across the slope (Fig. 4.3a). When a dam was established a pond was formed behind it, in which the bedload was deposited as a downslope thickening, microterrace, 1–3 cm thick (Fig. 4.2b) containing a variety of depositional crusts (Humphreys 1994b). The material as a whole had a coarser mean grain size, was better sorted and contained less silt/clay than the material from which it was derived (Table 4.1). The tendency for the differential removal of fines was accentuated during the waning stages of overland flow events, when rainsplash once again became important and left behind on the microterraces a capping of up to 5 mm of single-grain sand, from which the fines had been removed.

72

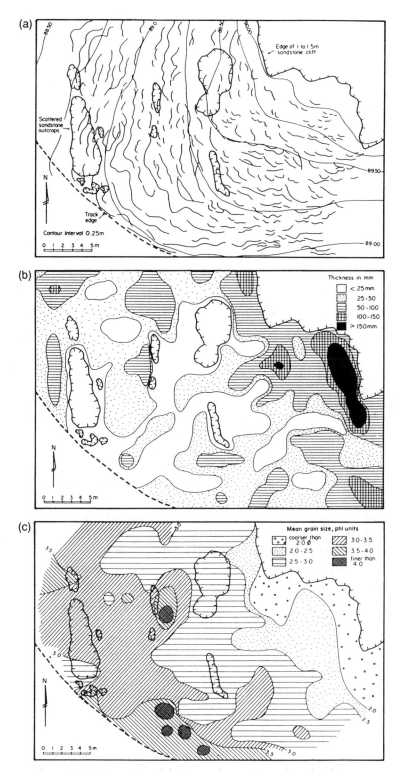

Figure 4.3 Depositional features of a sandstone bench (after Mitchell & Humphreys 1987): (a) litter dams; (b) topsoil thickness; (c) topsoil, mean grain size.

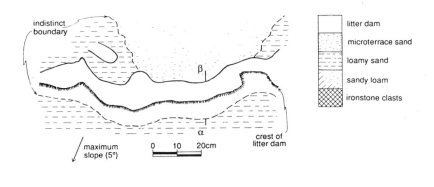

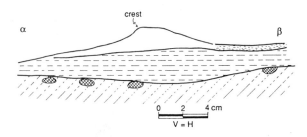

Figure 4.4 Litter dam (after Humphreys 1985).

Table 4.1 Sorting by rainwash (%; after Humphreys 1985).

	Sand (20–2000μm)	Silt (2–20μm)	Clay (>2μm)
Trapped sand on microterrace	96.8	1.5	1.7
Source material	88.0	7.7	4.3

Litter dams and microterraces were established on hillslopes within a very few days of a bushfire (Fig. 4.5a,b) and this immediately reduced the rate of sediment removal from hillslopes by rainwash, for a great deal of the potentially mobile material was temporarily stabilized. This process was further encouraged by the moisture concentrated within the microterrace sediments and responsible for the rapid growth of filamentous green algae and cyanobacteria, which increased erosion protection (Bond & Harris 1964) and stimulated the growth of longer-lasting vegetation (Fig. 4.2c,d). Observations, after the bushfires of January 1994 in the Sydney region, indicated that litter dams persisted in the landscape for 13 years at least, despite the intensity of the fires, to form a nucleus for the formation of new dams and microterraces. In other words, bushfires do not necessarily result in a completely new surface, for even in the near-surface zone there can be a degree of inheritance.

Figure 4.6 attempts to quantify the decrease in erosion losses more precisely, by plotting soil loss against time since the last fire (with time acting as a surrogate for ground cover). The sites all have similar topsoils derived

74

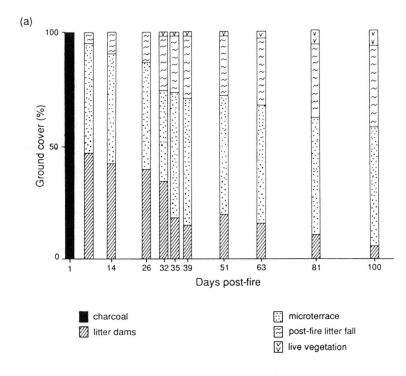

(a)

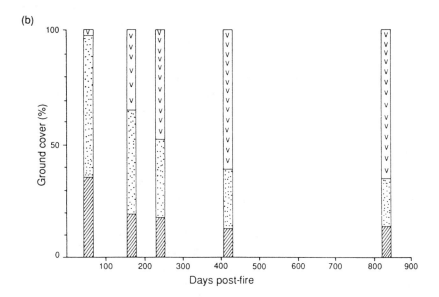

(b)

Figure 4.5 Post-fire changes in ground cover (after Mitchell and Humphreys 1987): (a) within 100 days; (b) within three years.

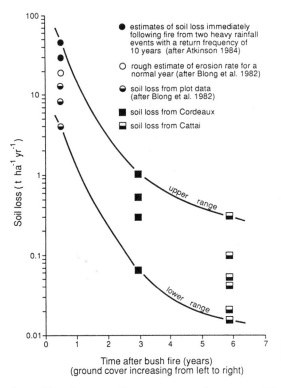

Figure 4.6 Soil loss after bushfire (after Humphreys 1985).

from Hawkesbury Sandstone and comparable ground slopes. The graph shows more than an order of magnitude decrease in soil loss after the first year, and therefore any longer-term average will be strongly influenced by fire frequency. During the first year after a fire soil loss by rainwash is likely to exceed the biological mounding rate. Over a five-year period the highest rates of average rainwash match the maximum rate of mounding. Thereafter, mounding exceeds rainwash, which is the "normal" situation, because in most environments average time between fires exceeds five years. Even with complete vegetation cover, however, erosion and downhill transport of soil does not cease altogether, for the mesofauna remain active and continue to deposit material on top of the litter layer, from where it can be easily mobilized. However, overall it can be taken that the removal of sediment from such hillslopes will be markedly periodic.

In attempting to compile an overall figure for sediment loss from hillslopes, account must be taken of mineral material rainsplashed onto the litter and retained, for it will then move down slope with the litter in the process of **rafting**. In the Ardennes, Imeson & van Zon (1980) and van Zon (1980) showed that rafted mineral material accounted for 24% of the total sediment delivered to the streams.

76

From the distribution of litter dams it can be concluded that they represent an important hillslope process, for they have been observed in undisturbed lowland savanna and in cultivated areas in the humid highlands of Papua New Guinea, in the tropical grasslands of the Northern Territory near Darwin and also in the semi-arid mulga woodlands near Tennant Creek, at many places on the tablelands and western slopes of the Great Dividing Range in New South Wales and Queensland, and in the subalpine grasslands of the Snowy Mountains (Mitchell & Humphreys 1987).

Litter dams are probably a worldwide phenomenon since they can be expected to form wherever conditions are suitable for the transport of copious amounts of organic debris over the soil surface. Thus, there is a considerable possibility that vegetation patterns of semi-arid areas in both Africa and Australia such as vegetation arcs (MacFadyen 1950a, Greenwood 1957, Hemming 1965), butana grass patterns (Worrall 1959), mulga groves (Slatyer 1961), contour banding in chenopod shrubland (Mabbutt 1972) and brousse tigrée (White 1970) may be expressions of litter dam phenomena on very low-angle slopes. However, such vegetation patterns are about two orders of magnitude larger than litter dams and in addition they vary in shape from being convex up slope across broad gentle drainage depressions, whereas those positioned along gentle hillslopes are convex down slope, as are all litter dams. On the other hand, the patterns are generally orientated parallel to the slope and perpendicular to any water flow and, although their precise origin has not been established, it is generally recognized that debris-transporting sheet floods or overland flow play a major part in their formation (MacFadyen 1950b, Greenwood 1957, White 1971).

Similar supportive evidence has more recently come from geomorphological work in both North America and Australia (Wells & Dohrenwend 1985, Patton et al. 1993) that identifies transverse bedforms on low-angle alluvial fans as resulting from high-magnitude sheet flow events. They bear a strong morphological resemblance to litter dams.

This discussion indicates strong parallels exist between the hydraulics of rainwash and river flow, in particular the capacity of shallow overland flow to transport and sort particulate matter (Emmett 1970). In this sense the fluvial system extends from river channels, to non-channelled hillslopes and up to the divides, even on very gentle slopes (Moss & Walker 1978).

It is now possible to consider the combined effects of bioturbation and rainwash on topsoil formation. Practically all the material deposited on the surface, either as mounds or as simple casts, comes from the topsoil and in general it is somewhat finer-grained than the topsoil. As a result of rainwash the material on the surface is further winnowed, with the fine fraction being preferentially removed in suspension and transported out of the system by

77

rivers and streams. This leaves behind the coarser dominantly quartz sand and the litter to move slowly and spasmodically downhill, often in the form of litter dams and microterraces. From time to time, as a result of bioturbation, this material is reincorporated into the topsoil and eventually, as a result of further bioturbation, it reappears once more on the surface to be subject to another cycle of winnowing. Because of the constant reincorporation of the surface winnowed material into the bioturbated layer, the effect of the winnowing penetrates down to the full depth of maximum bioturbation, that is it affects the whole of the topsoil. As a result the topsoil becomes coarser in texture and better sorted over time as it slowly descends the hillslope over the subjacent material, not *en masse* but grain by grain, and so leads to an ever more marked differentiation between topsoil and subsoil. The whole process is maintained by the mesofauna mining the top of the subsoil, which provides a continuous source of fines and causes the topsoil/subsoil boundary to sink into the landscape. The occasional occurrence of relic subsoil features at the base of the topsoil substantiates such boundary movement.

As was seen in Chapter 3 it is a natural consequence of bioturbation that clasts larger than the maximum size that can be manipulated by fauna will move downward to the limit of bioturbation to form a stone line (more correctly a stonelayer). This is also seen in Figure 3.5 where the stones accumulate at the base of the topsoil coincident with the disappearance of both tubules and relic tubules. However, bioturbation is only one element of stonelayer formation. The other results from the downhill movement of the topsoil. Figure 4.7 is a section of part of a gentle hillslope in the Brisbane Valley of southern Queensland, where a sandy loam topsoil overlies a sandy clay saprolitic subsoil that in turn grades into the underlying granite. The relationship of these two layers is made clear by the quartz vein, which is not affected within the saprolite, but is disrupted in the sandy loam surface and forms a mound just down slope from where it would outcrop. Down slope from the mound the quartz fragments occur as a distinct stonelayer at the junction between the topsoil and the saprolite. No stonelayer exists between the two layers up slope from the point of outcrop. This can be interpreted as showing that the sandy loam topsoil is moving down slope across the saprolite, which is being formed by *in situ* alteration of the granite (Plate 3).

A much more detailed investigation of stonelayers was made on the north side of Broken Bay at Patonga, New South Wales (Fig. 3.1), where the Hawkesbury Sandstone either outcropped as hard bedrock or formed a saprolite (Bishop et al. 1980). A well developed stonelayer occurred at the base of the topsoil (Fig. 4.8), which contained both equant and plate-like clasts. Bishop et al. (1980) investigated the plate-like clasts of the stonelayer

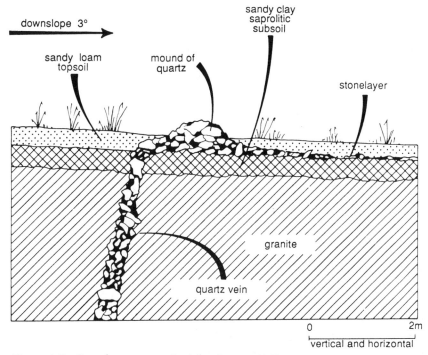

Figure 4.7   Stonelayer on granite (after Paton 1978).

at location H (Fig. 4.8) by removing the overlying finer-grain material and measuring the orientation and mutual relationships of all the plate-like clasts larger than 3 cm. It was clear that, whereas the slope of the top of the saprolite was 2°, all of the plate-like clasts had dips greater than this. These

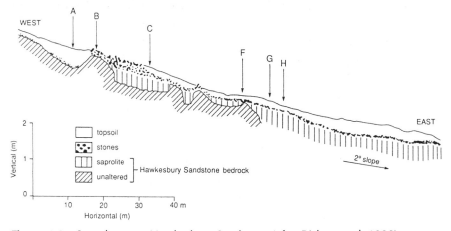

Figure 4.8   Stonelayer on Hawkesbury Sandstone (after Bishop et al. 1980).

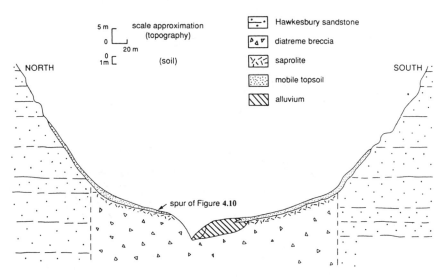

Figure 4.9 Soils of Peat's Crater (after Bishop et al. 1980).

dips were mostly down slope in the range 8° to 30°, but dips of up to 60° were not uncommon and a few were even higher. In addition, many of the plate-like clasts were in contact with one another in a stacked or imbricate fashion, which indicated a transportational origin for the stonelayer.

It was possible to pursue this analogy with sedimentary fabrics further, for within the topsoil both proximal and distal facies could be recognized. The proximal facies was a uniform mix of stones and finer-grain materials, which was located on, or slightly down slope from, sandstone benches that appeared to be the source for much of the material in the topsoil. The distal facies was characterized by having a well differentiated stonelayer in which the equant clasts became smaller and the ratio of the thickness of the topsoil to that of the stonelayer became greater the more distal its position. Thus, within the area covered by Figure 4.8 the proximal facies was between B and C and the distal facies down slope from F. At A the surface layer had the characteristics of an extreme distal facies, further indicating that the bench at B was the source of the material in the surface layer down slope from B. Collectively all of this evidence indicated the downhill mobility of the topsoil.

Rather more direct evidence of the downhill movement of the whole of the topsoil was apparent where Hawkesbury Sandstone had been penetrated by diatremes composed of brecciated basic igneous rocks (Bishop et al. 1980). Because they were eroded more easily than sandstone, the breccias occurred in depressions surrounded by steep slopes at the sandstone boundary (Fig. 4.9). Characteristic of the topsoil, no matter whether it was over the sandstone directly, or the saprolite, were quartz sand grains with well terminated pyramidal faces (Table 4.2) that were also common in the Hawkesbury Sandstone and yet were totally absent from the saprolite within the diatreme. Similar evidence was provided by heavy mineral separates of ilmenite, rutile and zircon, which also occurred in the topsoil and in the Hawkesbury Sandstone, but almost never in the saprolitic breccia (Table 4.2). This showed that the topsoil was derived from the Hawkesbury Sandstone, which had moved down slope transgressively across the saprolite, formed by the *in situ* epimorphism of the breccia.

The lower diatreme slopes also provided evidence of the differential lateral movement of fines within the topsoil. This was investigated on the north side of the depression where a gentle spur extended south across the sandstone/diatreme boundary towards the centre of the basin (Fig. 4.10). Down this spur towards the streamline there was a marked break in slope from 11° to 7° at the 36 m contour. This break in slope was also the point below which no pebble-size fragments of sandstone occurred. The surface layer was at its thickest (105 cm) immediately below this point and had a loamy sand texture. It then gradually decreased in thickness to 21 cm and increased in texture to a clay loam farther down the spur. Particle-size analysis of the surface layer showed that the decrease in median grain size along the ridge was directly related to the distance from the last sandstone outcrop and the overall decrease in slope. All of this supported the contention that the surface layer of soil material was mainly forming from the breakdown of Hawkesbury Sandstone up slope and was moving down

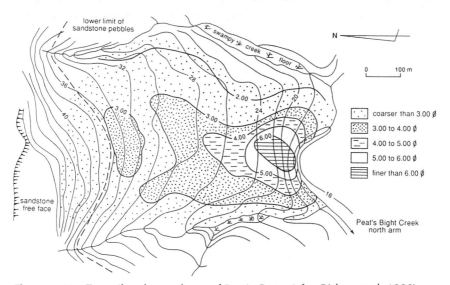

Figure 4.10 Topsoil on lower slopes of Peat's Crater (after Bishop et al. 1980).

81

across the diatreme boundary. In addition it demonstrated that bioturbation was operating in the saprolite sufficiently rapidly for it to "dilute" the sandstone-derived material on the more distal parts of the slope.

Table 4.2 Mineralogy of Peat's Crater (after Bishop et al. 1980).

	Topsoil	Saprolite	Sandstone	Breccia
Number of samples	8	8	3	3
Field texture	Sandy loam	Light clay	Rock	Rock
Sand (%)	80 (66–84)	13 (5–38)	72	–
Silt/clay (%)	20 (16–32)	87 (62–95)	28	–
Mean size (phi)	2.68	~8.00	1.77	–
Quartz crystal overgrowths	Abundant	Absent	Abundant	Absent
Heavy minerals (%)				
Rutile	20 (6–28)	1 (0–4)	21	0
Ilmenite	5 (0–6)	0 (0–2)	7	0
Zircon	10 (2–20)	0 (0–1)	15	0
Tourmaline	8 (0–14)	0 (0–2)	6	0
Leucoxene	13 (0–21)	0	37	0
Limonite	44 (24–87)	99 (94–100)	12	5
Microxenoliths	0	0	0	86
Xenocrysts	0	0	0	9
Others	0	0	2	0

The same processes of differential downslope movement of fines, with a coarse-grain mobile topsoil over a heavier-textured saprolitic subsoil (Fig. 4.11), were seen where granite was the bedrock at Killonbutta in central New South Wales (Mitchell 1985; Fig. 3.1). There was a topsoil up to 60 cm deep over both a fresh granite and a granitic saprolite. The mobile, transgressive nature of the topsoil was indicated by a general decrease in median grain size of particles down slope from the fresh granite outcrop. Thus, the microterraces produced by rainwash on the upper slope were

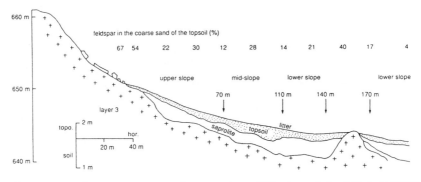

Figure 4.11 Texture contrast soil on granite at Killonbutta (after Mitchell 1985).

made up of coarse, compound fragments of disaggregated granite, which only persisted for 10–15 m down slope from the rock outcrop before they disintegrated into individual minerals. The feldspars freed in this process at this point made up 67% of the coarse sand fraction, which decreased to only 4% on the lower slopes as a result of epimorphism during slow downhill movement, which was also responsible for a reduction in mean particle size from 0.41 mm to 0.09 mm. This downslope movement was reflected in the microterraces of the lower slopes having a very fine sand texture, compared to the much coarser materials of their counterparts on the upper slope.

Up to this point discussion has been restricted to materials that were heterogenous enough to produce texture-contrast soils when affected by bioturbation and rainwash. However, there were situations where the material produced by epimorphism, or directly inherited from bedrock, was uniformly fine-grain, so there was nothing to be sorted by bioturbation, or surface winnowing. Nevertheless, bioturbation and near-surface processes were still operating, so that a transgressive topsoil was produced, which was differentiated from the underlying saprolite not in terms of texture, but rather in terms of fabric. Such a soil was developed on dolerite at Prospect near Sydney (Fig. 3.1, 4.12; Hart 1988). The dolerite, which formed the hill, was intrusive into a series of sandstones and shales. In a midslope position it was possible to differentiate a topsoil 20–30 cm thick from a saprolite of about 40 cm thickness that graded into the dolerite. Both topsoil and saprolite had clay textures, but they differed in their fabrics. This was well seen in the case of void development, for in the topsoil the voids had a maximum elongation at 60–75°, which was very close to the slope azimuth (55°), whereas in the saprolite the maximum elongation was at 165–180°. The topsoil voids were for the most part associated with the packing of

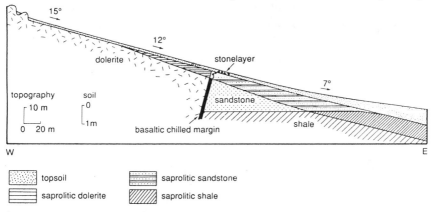

Figure 4.12 Fabric-contrast soils on dolerite at Prospect (after Hart 1988): cross section of Prospect dolerite margin.

faecal pellets within worm channels. The saprolite voids, however, were inherited from the joint planes in the dolerite (Hart et al. 1985) and this gave rise to strong blocky peds. Worm activity and plant roots in this case were confined to the planar voids and did not penetrate within the peds. Up slope the saprolite died out and the topsoil was directly over the dolerite. Down slope the boundary of the intrusion was very narrow and marked by a fine-grain chilled margin. Immediately beyond that boundary there was no more dolerite-derived saprolite, but one formed from sandstone. The dolerite-derived topsoil, however, continued across this boundary without any fundamental change in its character for a distance of about 40 m, before coarser-grain particles from the sandstone became apparent. At the boundary itself, and for a distance of about 20 m down slope from it, a stonelayer was developed at the base of the topsoil where the clasts consisted of the chilled margin of the intrusive, which because of its fine-grain nature was more resistant to epimorphism. It was clear that in this instance a mobile topsoil was being dealt with, which was transgressive across a varied saprolitic substrate. Where this substrate was derived from dolerite and had a clay texture, a **fabric-contrast soil** was developed.

A more widespread development of fabric-contrast soils was found to occur on the Darling Downs of southern Queensland (Paton 1974). In general the landscape consisted of flat-topped basaltic residuals surrounded by long gentle slopes, at the upper end of which the topsoil was a shallow gravelly clay, which transgressed a varied sequence of basalts without any alteration in its character (Fig. 4.13a). However, examination of the gravel content of the topsoil showed that each of these different basaltic materials gave rise to a different type of gravel and each of them could be traced for a short distance down slope from its point of outcrop. This was well seen in the case of the red bole. The fact that the gravels could be followed only in a downslope direction was strong evidence of downslope movement of the topsoil. Furthermore, it can be inferred that such movement was at a slow rate, for, although it must have taken some time for these gravels to lose their distinctive character, they did so within a few metres of their point of origin. Farther down slope (Fig. 4.13b) the topsoil thickened and transgressed basaltic saprolite to form fabric-contrast soils.

Conclusion

By combining the data presented in this chapter with those in Chapters 1–3 it is now possible to derive an overall model of soil formation on hillslopes, which involves the epimorphism of bedrock to form saprolite, its differential

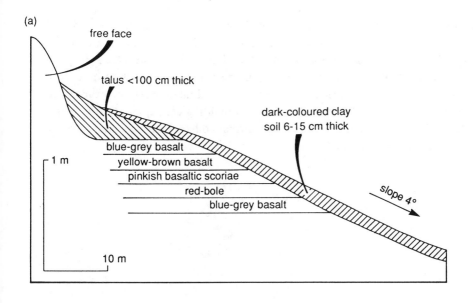

(a) free face

talus <100 cm thick

dark-coloured clay
soil 6-15 cm thick

1 m

blue-grey basalt
yellow-brown basalt
pinkish basaltic scoriae
red-bole
blue-grey basalt

slope 4°

10 m

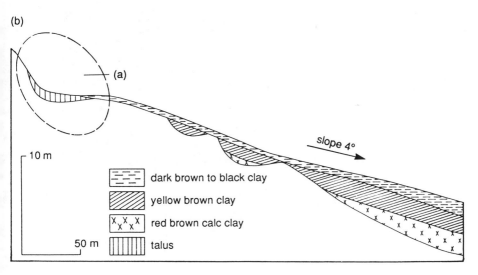

(b)

(a)

slope 4°

10 m

	dark brown to black clay
	yellow brown clay
ˣ ˣ ˣ ˣ	red brown calc clay
	talus

50 m

Figure 4.13 Fabric-contrast soils Darling Downs (after Paton 1978):
(a) upper slopes; (b) lower slopes.

mining by mesofauna to form a topsoil, which is then further sorted and moved down slope by rainwash. This results in the formation of a contrast soil, which consists of a mobile biomantle, often with a stonelayer at its base over a subsoil saprolite. For the most part these contrast soils have a texture contrast (i.e. the biomantle is dominated by residual quartz); only rarely are they fabric contrast. In other words a full pedological explanation is being offered for the formation of texture-contrast soils, which was shown to be impossible from a zonalist base.

1 Mound of *Aphaenogaster longiceps* (see Ch. 3), Sydney region.

2   Litter dam and microterrace (see Ch. 4), Sydney region.

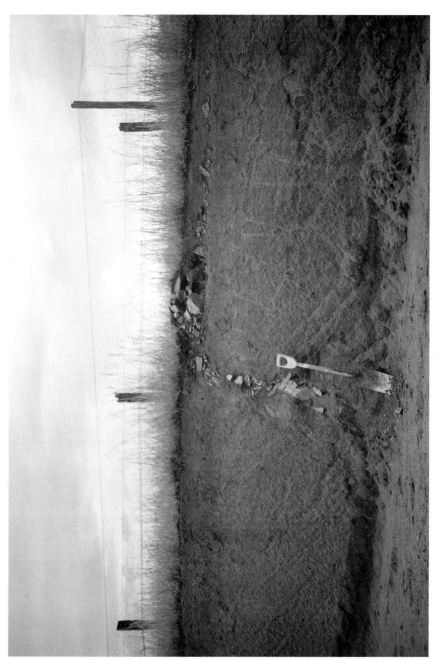

3 Stonelayer (see Fig. 4.7), southeast Queensland.

4 Mukkara developed on Tertiary basalts (see Fig 6.6 and Ch. 6), southeast Queensland.

5 Fabric-contrast soil where a highly pedal dark clay topsoil transgresses three very different subsoils (see Ch. 8), southeast Queensland.

6 Podzol developed in a deep uniform sand (see Ch. 8), Sydney region.

7 Texture-contrast soil with thick topsoil, a stonelayer and a very deep saprolite (see Ch. 10), Cameron Highlands, Malaysia.

8 Fabric-contrast soil with the base of the topsoil being marked by a stonelayer and the underlying saprolite by a perched gley (see Ch. 10), southern Thailand.

CHAPTER 5

Aeolian processes

About thirty years ago there was much talk that geologists ought only to observe and not theorize; and I remember someone saying that at this rate a man might as well go into a gravel pit and count the pebbles and describe the colours. How odd it is that anyone should not see that all observations must be for or against some view if it is to be of any service. *Charles Darwin*

So far the detachment and consequent transport, sorting and deposition of lithospheric material in particulate form by water has been discussed. The only other agent that acts on individual particles is wind. However, its power is much less than that of water and this restricts its realm of operation to environments where there is little vegetation and an abundance of loose surface materials. To a large extent this means deserts, both hot and cold, as well as more confined areas along coasts and lake shores, and occasionally after vegetation destruction by fire or drought.

Within such areas the kind of aeolian transport and the resulting deposits are strongly influenced by the availability of materials of suitable particle size. Generally it is unusual for particles coarser than sand size (>2 mm diameter) to be moved by wind across a rough surface. However, particles that are finer than this, but coarser than 60–75 μm (i.e. sand size), can be moved by saltation over distances of up to several kilometres, to form sand sheets and cover sands in which various dune forms may develop (Mabbutt 1977). Even so, it needs to be emphasized that it is the particle size of the transported and deposited material that is being discussed, which may differ from the ultimate particle size involved. Thus, clay dunes (Bowler 1973) or lunettes, which occur in semi-arid areas of southeastern Australia on the downwind side of seasonally dry lakes, are not formed by the mobilization and deposition of individual clay platelets, but rather by the wind erosion of mud curls and salt mud efflorescences on dry lake beds, which in the process become rounded and compacted largely as sand-size pellets. As such these pellets move by saltation and accumulate down wind, in the same way as any other sand-size particles, the coarser pellets as low dunes and the finer ones as thin discontinuous **parna** sheets (Butler 1956). Thus, despite being formed of clay particles, their accumulation as discrete depositional bodies is a reflection of their movement as coarser particles (Dare-Edwards 1984).

It is only when particles are finer than 60–75 μm in diameter (i.e. dust) that transport in suspension becomes possible. However, for such movement to be initiated the prior saltatory impact of coarser sand-size particles is required. Therefore, the source material, as well as occurring in areas of limited vegetation, must have a non-uniform grain size. Such a concurrence of favourable factors can lead to the transport of dust particles over great distances (Pye 1987). At present such regions of potential dust export are relatively restricted, and from Figure 5.1 it is apparent that the greater part

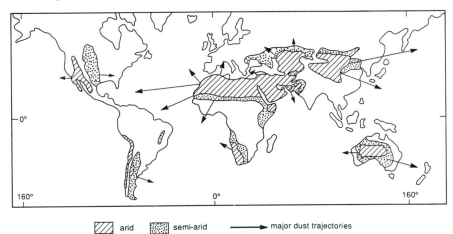

| ⟋⟋ arid | ⠿ semi-arid | ⟶ major dust trajectories |

Figure 5.1   Dust storm trajectories (after Pye 1987).

of the dust must be deposited directly into the oceans and only in the case of West Africa, southeast Australia, the Mediterranean littoral, southern USSR and China would terrestrial deposits be important.

In the previous chapter it was seen that the overall effect of rainwash and streamflow was to remove the fines preferentially from surface soil as suspended load to sedimentary sinks in rivers or oceans, leaving behind the coarser sands on the continents. A common feature, however, of both Africa and Australia is not of through-drainage, but of basins of inland drainage, which means that the fine suspended load of the streams is deposited within such basins. The Lake Chad Basin in West Africa (McTainsh 1985) spans 15° of latitude, from the equatorial conditions of the Congo watershed in the south to the Sahara Desert in the north (Fig. 5.2). The area is drained by the Logone and Chari Rivers, which carry fine-grain suspended sediment to near the centre of the basin. In the same area are mobile dunefields of

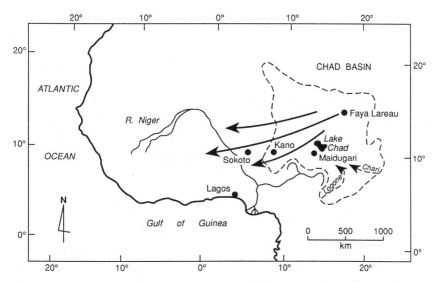

Figure 5.2 Harmattan dust transport in West Africa (after McTainsh 1985).

coarse-grain sand, the saltatory movement of which provides the trigger for the entrainment of the river-deposited fines as dust clouds. This is the result of the *harmattan*, which blows periodically to the west and southwest from October to April and carries with it the dust that originated in the Chad Basin over the West African savanna lands and the Atlantic Ocean, where it is gradually deposited (Prospero & Carlson 1972, Morales 1979, Prospero et al. 1981). Both the rate of dust deposition and the particle size of the deposited dust decreases down wind. The nature of these deposits was examined in the Kano area of northern Nigeria (McTainsh 1984), and this showed that on average dust deposition in a year would be equivalent to a layer 74 μm thick. Particle-size analyses, with peaks at 44 μm and 75 μm, showed that dust additions are dominant to a depth of up to 50 cm on high parts of the landscape.

A similar conjunction of circumstances occurs in central Australia (McTainsh 1985) where the Lake Eyre Basin encompasses a great range of environments, from the seasonally humid tropics of Queensland, to the arid centre (Fig. 5.3). The south-flowing Eyre and Diamantina Rivers and Coopers Creek, with their extremely low gradients, deposit their finer sediments in arid areas as the water dissipates *en route* to Lake Eyre. Such fine-grain sediments are entrained in dust clouds where the south-flowing river systems are traversed by the mobile dune sands of the Simpson Desert. This provides the necessary trigger mechanism under the influence of strong winds associated with the passage of frontal weather systems across the central Lake Eyre Basin (Sprigg 1982; Fig. 5.3) and explains the two major

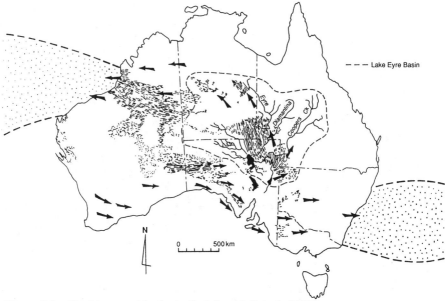

Figure 5.3 Dust transport in Australia (after McTainsh 1985).

dust-transporting systems, one to the northwest and one to the southeast. The Australian system is, however, presumably much weaker than the West African/Saharan system, and this is reflected in the less than conclusive evidence that is available on aeolian mantles in Australia. An investigation of the desert loams in the Barrier Range to the north of Broken Hill, New South Wales (Chartres 1982; Fig. 3.1), is informative for the site lies in the middle of the dust-transporting system. It was concluded that much of the silt and clay within these desert loams, which are about 70 cm deep, was not derived from the bedrock, but was aeolian in origin. However, it was not possible to distinguish the aeolian contribution quite as clearly as was done in West Africa.

The greater part of this fallout from dust plumes occurs within a few hundred kilometres of them being generated, so that in cases such as those of Lake Chad and Lake Eyre most of the dust is deposited within these basins of internal drainage. In terms of the mass of material involved it is possible to regard the movement of these fine-grain materials as a closed cycle, for the bulk of the aeolian dust deposited within the basins is moved back towards the arid centre as suspended load in streams, where it is redeposited, ready for entrainment again in dust plumes. Nevertheless, there is some "leakage" of dust across the drainage divides into more humid areas, such as the forested coastal zone of West Africa and the coastal zone of southeastern Australia. However, by this stage deposition from the dust plume is spread over a much greater area, so that the necessarily much smaller amount of dust deposited on any given area is more difficult to recognize, particularly in view of the greater impact of rainwash and bioturbation, as compared to more arid areas.

There are, however, sites where a stark compositional contrast between bedrock and aeolian material renders recognition much easier. At Belarabon in western New South Wales (Fig. 3.1) valley fills contain carbonates and clay minerals, which could not have been derived from the local siliceous bedrock and are interpreted as being reworked aeolian dust (Wasson 1982). Dust mantles up to 20 mm thick of clay- and silt-size quartz are also found trapped in moss and lichen mats on rock outcrops all along the western slopes of the Great Dividing Range of New South Wales. The Negev of Israel provides another example where dust is accumulating over various bedrocks (Yaalon & Dan 1974). At one site the dust consists of clay minerals and grains of quartz, calcite and dolomite, which is totally different from the mineral composition of the underlying diabase (Dunin & Ganor 1991). A much more important example of possible aeolian dust addition is provided by the *terra-rossas* of the Mediterranean. These fine-grain red soils, up to 50 cm deep, overlie hard, very pure limestones across a sharp boundary and are widespread in Greece, Italy, southern France and Spain. Two main theories have been used to explain their formation: that they are the insoluble residue of the limestone, or that they result from the deposition of aeolian dust carried from the North African deserts by the *sirocco*. It has now been established from work in Greece (Macleod 1980) that there is no correspondence between the particle-size distribution of limestone solution residues and that of the *terra-rossa* soils, which is more akin to that of aeolian dusts (2–75 μm). As a consequence it is now more generally accepted that these soils have an aeolian origin.

As long as the strong mineralogical and particle-size contrast between aeolian dust accessions and the bedrock is maintained, it is possible to differentiate between them, even on oceanic islands, far away from any immediate and obvious dust source, where total aeolian addition is very small. On the Hawaiian islands, Jackson et al. (1971) (see Ch. 2) discovered some very fine, fragmental, angular quartz, 70% of which was in the 2–10 μm range, in the surface horizons of soils, supposedly derived by epimorphism from quartz-free basic rock. Detailed investigations of this quartz in all the major islands demonstrated that the amount varied directly with the rainfall and also was affected by the source of the rainfall, as well as with elevation and age of landscape. This is strong evidence for the quartz arriving as an aeolian dust to be rained out on the islands; a conclusion supported by the occurrence of exactly the same type of quartz in deep-sea cores from both north and south of the islands. Confirmatory evidence for these conclusions was derived from the oxygen isotope ratios of quartz, which were quite low and did not vary a great deal. This pointed to a very well mixed continental source and definitely eliminated any suggestion of low-temperature near-surface neogenesis.

Yet further support for aeolian additions to Hawaiian soils came from an equally detailed study of fine-grain potassium micas, which also occurred in the surface layer of these soils and could not be explained as resulting from the epimorphism of the bedrock. Initially their presence was explained by the surface enrichment of potassium by plants, which created conditions suitable for their formation (Juang & Uehara 1968). However, considerable doubt developed about such an interpretation, for micas newly formed under such conditions should have had an acicular rather than an equidimensional habit. Furthermore, the fine grain size of the micas and the fact that they co-varied in association with the quartz pointed to a common aeolian origin. This conclusion was confirmed when ^{40}K/^{40}Ar dating gave the mica an age of 200 Myr as compared to 3.5 Myr for the age of Oahu from which the mica sample was obtained (Dymond et al. 1974).

The pedological importance of wind in terms of detachment, transport, sorting and deposition of soil material is particularly well seen in those semi-arid areas where the balance between the vegetation cover, bioturbation, rainwash and aeolian activity is more delicately poised. Such a situation was investigated (Goldrick 1990) on the alluvial plains just to the west of Condobolin in western New South Wales (Fig. 3.1) where the average rainfall was 400–450 mm yr^{-1}, with extremes of 195 mm and 827 mm yr^{-1}, and the vegetation consisted of a mixture of low open forest and tussock grassland. The greater part of the area investigated consisted of Quaternary prior-stream deposits where the relief rarely exceeded 1 m, although there were occasional outcrops of steeply dipping metamorphosed Palaeozoic sediments, where the relief was up to 20 m. On both of these materials well marked texture-contrast soils developed with typical red-brown earth and solodized-solonetz morphology. In both cases there was a sufficient range of particle size in the underlying material to form a texture-contrast soil, and, in addition, the Palaeozoic bedrock contained sufficient quartzite fragments to form a stonelayer. The abundant evidence of mesofaunal activity in the form of both surface casts and subsoil burrows, the presence of bedding in parts of the topsoil unrelated to the bedding structures of the underlying subsoil or bedrock, and the transgressive nature of the topsoil across the subsoil clay, all pointed to the topsoil being derived from subsoil material by a combination of bioturbation and the sorting action of rainwash and wind. This moved material laterally down slope and down wind, and removed the fines by winnowing to leave the coarser sands behind. The coarse texture of the topsoil, together with its single-grain or highly porous earthy-sand fabric, allowed rainfall to penetrate to the topsoil/saprolite boundary with ease. This caused a certain amount of dispersion of the saprolitic clay along the interface and down into the vertical planar voids of the saprolite, creating as a result domed columns in the subsoil (Fig. 5.4a). The clay moved only a short way down the planar voids before it was deposited, which caused the base of the topsoil to become waterlogged and bleached; temporary induration followed subsequent evaporation. However, much of the area had been extensively overgrazed and this has resulted in a patchy ground cover with a mosaic of tussock grasses and forbs, which varied in both space and time depending on species composition, seasonal conditions, grazing intensity and the nature and depth of the soil. To cover such a range of conditions it was necessary to describe the balance between the various processes under dense and sparse vegetation cover. In addition, two intermediate situations of medium cover had to be considered, for the paths to and from the vegetational extremes differed from one another, rather like the two arms of a hysteresis curve in which one involved degradation of the topsoil (Fig. 5.4a–c) and the other its restoration (Fig. 5.4c–a).

Under dense vegetation cover the soil had a stable level surface, for lateral movement was minimized and effects of bioturbation maximized so that the small amount of fines-depleted material produced by minimal surface winnowing was rapidly incorporated. Medium but decreasing vegetation exposed more surface casts to the impact of saltating sand particles and raindrops. As a result of the first, fines were removed in aeolian suspension, whereas those resulting from raindrop impact were removed by rainwash to be redeposited on lower areas as surface crusts or removed completely. The effects of this winnowing were naturally concentrated in the non-vegetated areas, where the topsoil that remained was of a coarser texture and more single-grain. Sparse vegetation cover allowed wind erosion to become dominant, so that most of the topsoil was removed by deflation, except for isolated islands protected by vegetation. The limit to wind erosion was set by the indurated basal topsoil, which was exposed as cemented sand scalds. However, given sufficient rain, such features softened and dispersed under raindrop impact, to be removed by rainwash, which exposed the top of the subsoil forming a clay scald. Medium but increasing vegetational cover resulted from increasing rainfall, and the erosive process continued until the clay scalds were sufficiently broad and flat to pond for extended periods. The products of wind erosion were intercepted by patches of remnant and increasing vegetation to form topsoil hummocks adjacent to the clay scalds. At the same time vegetation was initiated on the clay scalds where ponding was most prolonged and this in turn caused topsoil accumulation. This

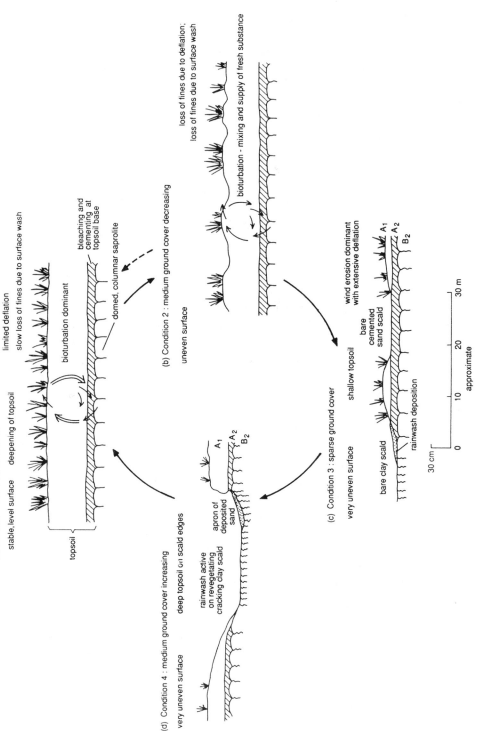

Figure 5.4 The balance of process in a texture-contrast soil of a semi-arid region (after Goldrick 1990).

general addition of topsoil as a result of aeolian and rainwash processes resulted in a reversion towards those conditions found under dense vegetational cover. On the soils derived from the Palaeozoic rocks, the stonelayer left behind by wind erosion as a residual pavement on top of the saprolite would be buried at this stage by wind-deposited topsoil to become a stonelayer once again. However, it needs to be emphasized that the stonelayer has not changed its position in relation to the saprolite since it was originally formed at the base of the topsoil by bioturbation and in its present situation it is inherited.

Hence it is apparent that in semi-arid areas a very sensitive balance exists between the pedogenic processes controlled by wind and water. Over distances as little as 10 m small differences in ground cover, bioturbation and the efficacy of wind and water erosion (Fig. 5.4) resulted in substantial variations in the thickness and extent of the biomantle, and it is for this reason that such areas are susceptible to land degradation.

Conclusion

At the scale of the large basins of internal drainage the closed cycling of dust has been emphasized. However, it would seem that such a system is probably very sensitive to regional environmental change, which could profoundly affect the dynamics of topsoil development across a very wide area.

In overall terms the main aeolian effect from a pedological point of view is to aid in the development of texture-contrast soils by winnowing fines and leaving behind coarser well sorted topsoils, which are capable of forming mobile sand dunes. In other words the wind is responsible for the accumulation of quartz sand as the final residual product of pedogenesis, which is such a feature of both Australia and Africa.

At the beginning of this chapter it was pointed out that at the present time source areas for dust and winds for the generation of plumes are not very common. However, during the Quaternary, with its repeated glaciations, source areas for dust as well as stronger winds were much more common and resulted in much more important aeolian deposits, such as the mid-latitude loess (Catt 1986). This of course gives to such materials a strong element of inheritance and will be discussed at greater length in Chapter 10. In addition, wind influences the distribution of the products of explosive volcanic eruptions and this will be considered in Chapter 11.

CHAPTER 6

Soil creep

The man who is certain he is right is almost sure to be wrong; and he has the additional misfortune of inevitably remaining so. All theories are fixed upon uncertain data and all of them want alteration and support. *Michael Faraday*

The near-surface processes discussed in Chapters 3, 4 and 5 dealt with the movement of materials as particles. Very different is the slow, gravity-induced, downslope mass movement of soil material generally referred to as **creep**. This was first invoked by Davis (1892) to explain summit convexities and ever since it has been an integral and important part of geomorphology. Despite this long history soil creep remains a highly equivocal concept (Finlayson 1985) that in large part derives from the original vagueness of definition. Sharpe (1938) believed that evidence of soil creep on hillslopes was widespread and easy to recognize (Fig. 6.1). Since then practically all this evidence has been refuted. Thus, tree curvature (B) is explained best in terms of plant growth (Phipps 1974), and the displacement of manmade objects (D,E,F) results from the direct imposition of an extra load over a relatively small surface area, rather than by the indirect effect of gravity (Statham 1977). In addition, it has been shown that stonelayers at the base of the topsoil (H) are much more cogently explicable in terms of bioturbation, rainwash and aeolian action, rather than resulting from creep. Given the efficacy of these other soil-transporting mechanisms it can be concluded that their combined effects would exceed any that result from creep by a considerable margin, so that attempts to measure creep in topsoils in these circumstances is labour in vain.

The results of slow mass movement have been observed in saprolitic subsoils. Figure 6.2 shows how the orientation of bladed ironstone clasts changed from being parallel to the bedding at the base of the subsoil towards the direction of maximum topographic slope at its top. This slight but undoubted directional change can be compared to the more chaotic arrangement of the same type of clasts in the topsoil, which reflects the impact of bioturbation. Such subsoil movements have been referred to as

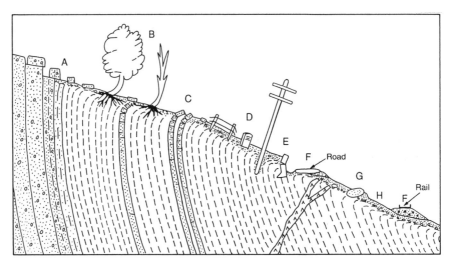

Figure 6.1 Soil creep (after Sharpe 1938). Common evidences of creep: (A) moved joint blocks; (B) trees with curved trunks concave up slope; (C) downslope bending and drag of bedded rock, weathered veins, etc., also present beneath soil elsewhere on the slope; (D) displaced posts, poles and monuments; (E) broken or displaced retaining walls and foundations; (F) roads and railroads moved out of alignment; (G) turf rolls down slope from creeping boulders; (H) stone line at approximate base of creeping soil.

creep, even though it has not been established that gravity is the trigger mechanism. A more likely explanation is in terms of rheid flow in response to differential loading, which has long been recognized in geology (Carey 1954). As well as controlling the formation of large features such as salt domes (Holmes 1965), its effects are evident at a much smaller scale in many near-surface materials. Thus, McCallien et al. (1960) described structures on a broad flat spur 30 m above sea level at Accra, Ghana. Four layers were distinguished above the basement gneiss, which from the surface down were: sandy loam, pisolitic ironstone, quartz breccia and heavy clay (Fig. 6.3). The structures were explained by a localized thickening of the quartz breccia and the pisolitic ironstone, which caused differential loading of the underlying clay. As a result the clay moved from areas of high over-burden pressure to areas of low overburden pressure, which eventually caused the clay to burst through the overlying layers and spread over the surface. Such piercement structures are referred to as **diapirs**.

The mudlump islands of the Mississippi Delta provide another example from less consolidated sediments. Mudlumps (Fig. 6.4) are actively developing clay masses formed within the main channels of the Mississippi Delta (Morgan et al. 1968) as a result of the bar sediments of the advancing delta mouths exerting unequal pressures on the underlying pro-delta clays. As a

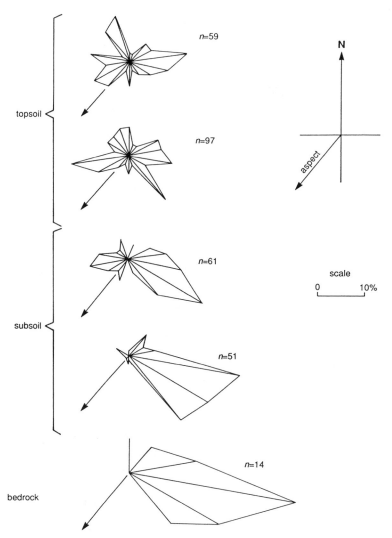

Figure 6.2 Subsoil movement.

result the clays move laterally and intrude into and through overlying materials to reach the surface as mudlumps.

Very similar structures also ascribable to differential loading are common in certain soil materials. Thus, within broad stretches of alluvial fill where the individual beds of material have a markedly variable thickness, the resulting differential overburden pressure leads to rheid flow and the development of piercement structures (**mukkara**; Paton 1974). Figure 6.5 shows a typical example from the Bremer Valley 50 km west of Brisbane in which finely pedal calcareous clays penetrate upwards and through a

99

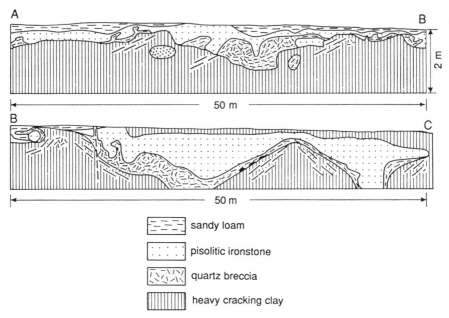

sandy loam

pisolitic ironstone

quartz breccia

heavy cracking clay

Figure 6.3 Piercement structure, Accra (after McCallien et al. 1960).

coarsely pedal dense clay, in a manner very like that of a mudlump, but on a very much smaller scale.

In certain cases bedrock is also involved in these mukkara. Figure 6.6 is a section of a 4° hillslope across gently dipping Tertiary basalts in the Brisbane region. Three flows are shown; the upper and lower one have been altered to a brownish yellow saprolite, whereas the middle flow, which has been subject to intense late-stage hydrothermal alteration, is a mixture of calcium carbonate and smectite clay (see Ch. 1). There is a surface layer, 20–30 cm thick, of a dense, dark, coarsely structured clay that is moving down slope. Immediately this encounters the calcareous, smectite clay mix, the latter flows and forms mukkara through the surface clay layer. Another example of bedrock being involved in mukkara comes from the same area (Fig. 6.7). In this case Tertiary basalts overlie a Mesozoic calcareous siltstone. Spreading outwards from the basalt cap is a series of stream-like hollows almost completely filled with a dark, coarsely pedal, dense clay derived from the basalt. Figure 6.7a is a cross section taken about 40 m away from and 6 m lower than the basalt cap. On the surface there are two depressions about 1 m deep and 12 m wide, within which there is a dark clay infill up to 1.5 m thick. In this case it is suggested that a dark clay infill is moving down slope and in doing so exerts maximum loading near the centre of the depression on the bedrock. Minimum loading would occur on the shoulders of the depressions and it is at this point that a mukkara of the type shown in Figure 6.7b is developed, which has a core of calcareous siltstone bedrock.

The clearest example showing the relationship between bedrock, soil material, topography and mukkara is provided from a site on the Darling Downs (Paton 1974) (Fig. 6.8; see also Ch. 2 and Fig. 4.13a,b). The landscape consists of flat-topped basaltic residuals, which are surrounded by an escarpment or debris slope, 3–12 m high, and then by 3–7° slopes up to 2 km long. The basaltic bedrock is highly variable consisting of blue-grey lavas and their reddish and yellowish brown alteration products, pale red and blue-grey scoria as well as red bole (Fig. 4.13a). A coarsely pedal, dense, dark-coloured clay forms the surface throughout and varies in thickness from 15 cm at the base of the debris slope to nearly 2 m near the drainage lines. In the middle and lower parts of the hillslope this surface layer is

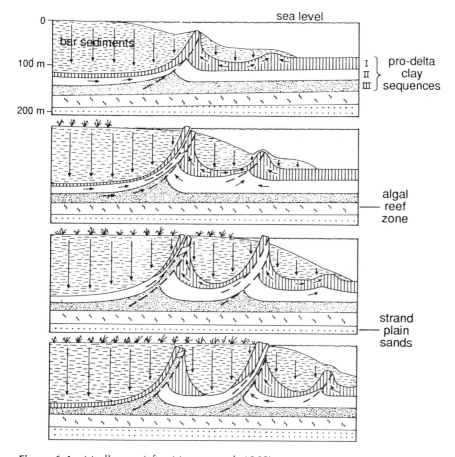

Figure 6.4 Mudlumps (after Morgan et al. 1968).

101

underlain by saprolitic yellowish brown and reddish brown, finely pedal, highly calcareous clay (Fig. 4.13b). Mukkara of various types are developed throughout this landscape and can be ascribed to the downhill mobility of the surface layer, evidence for which has been discussed previously. This movement leads to a "streaming" of the surface layer, similar to that which develops on the surface of flowing basalt lava, and the resulting differential pressure on the subsurface gives rise to mukkara in the intervening areas of lower pressure, which takes on a linear form down the hillslope. Naturally these mukkara are best developed where the highly calcareous subsoil clay occurs, but they are also to be seen in incipient forms on the upper slope, where they are derived from the more susceptible types of bedrock to form a vertical line of gravel. This finely developed linear form continues to the lower part of the hillslope, where the hillslope surface layer of dark clay

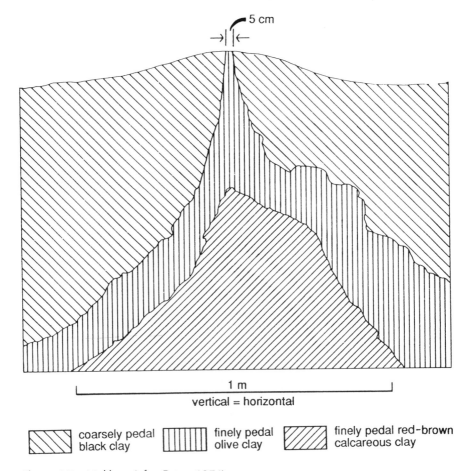

Figure 6.5 Mukkara (after Paton 1974).

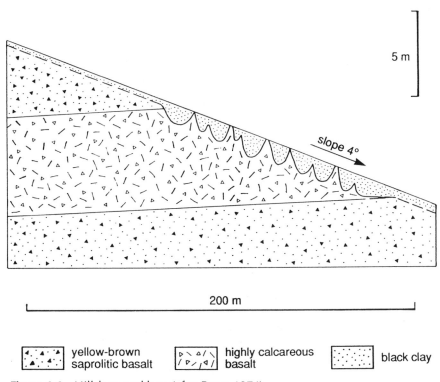

5 m

slope 4°

200 m

| yellow-brown saprolitic basalt | highly calcareous basalt | black clay |

Figure 6.6 Hillslope mukkara (after Paton 1974).

meets the much slower-moving clays of the valley floor, and the relatively simple pressure loading relationships are distorted. This is reflected in the shapes of the mukkara, some of which take on a scroll-like form and others in which the top of the scroll has been completely detached from its roots and carried several metres down slope, rather like nappe detachment in Alpine tectonics. On the valley floor itself the response to the loading coming from the mobile hillslope material is a more general one and a pattern of mukkara are developed, which are much greater in cross-sectional area than are those of the hillslope.

This form of subsoil flow is obviously of great importance, particularly in clay soils, but the significance of such subsoil movement has been rather obscured by the fact that the formation of mukkara is often accompanied by that of surface mounds and depressions, to which the term **gilgai** has been applied. Most explanations of gilgai have concentrated on the surface form and processes operating from the surface downwards (Hallsworth et al. 1955, Hallsworth & Beckmann 1969). The alternative suggested here is that the fundamental mechanism is load-induced movement of subsoil,

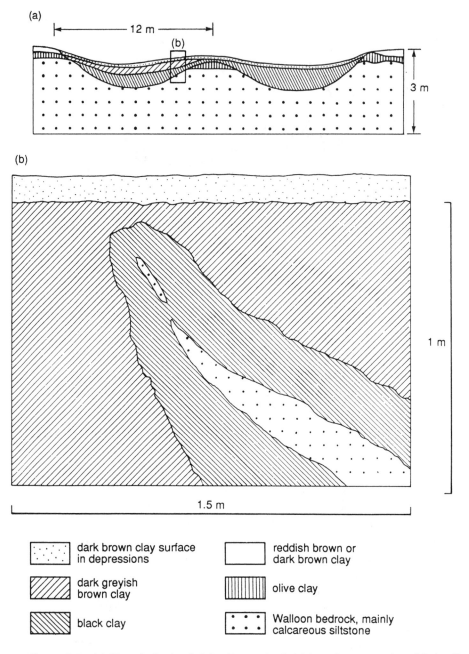

(a)

12 m

(b)

3 m

(b)

1 m

1.5 m

dark brown clay surface
in depressions

reddish brown or
dark brown clay

dark greyish
brown clay

olive clay

black clay

Walloon bedrock, mainly
calcareous siltstone

Figure 6.7 Mukkara in bedrock (after Paton 1974): (a) broad cross section; (b) detail.

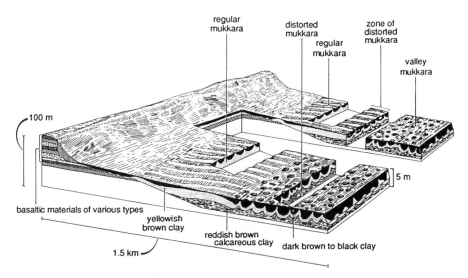

regular
mukkara

distorted
mukkara

zone of
distorted
mukkara

regular
mukkara

valley
mukkara

100 m

5 m

basaltic materials of various types
yellowish
brown clay
reddish brown
calcareous clay   dark brown to black clay
1.5 km

Figure 6.8 Mukkara, Darling Downs (after Paton 1974).

leading to mukkara development, with gilgai development being a super-
ficiality, which may or may not occur, depending on the rate of topsoil
mobility.

Another thing that continues to cause problems is the term "gilgai" itself,
for it has been applied to any form of micro-relief in clay soils, which
includes forms that are not associated with subsoil mukkara in any way,
such as the great variety produced by faunal activity, or others that result
from subsoil sapping.

Conclusion

This chapter has been restricted to a consideration of creep, or slow mass
movement of soil material resulting from the action of gravity. It has been
argued that it is essentially impossible to detect such movement in the topsoil, or biomantle, because of the speed of action of other processes such as
rainwash and bioturbation. Although slow mass movement is a common
phenomenon in clay-rich saprolitic subsoils and in complex alluvial deposits, it is largely explicable in terms of rheid flow resulting from differential
loading.

105

PART II

THE DISTRIBUTION OF SOIL MATERIAL

CHAPTER 7

The pedological hierarchy

My tactic is to make sweeping categorical statements. Whether or not this is a fault, in the free world of interchange of scientific ideas, is debatable. My own feeling is that it leads more quickly to the ultimate solution of scientific problems than a cautious sitting on the fence. *Ernst Mayr*

In the Introduction it was shown that the continued, but largely unconscious, adherence to soil zonalism gave rise to several problems, which in Part I were found to be resolvable by taking a much broader view of what was accepted as pedogenic processes. Fundamental to such a viewpoint was the realization that both epimorphic and near-surface processes had to be taken into account and yet zonalism excluded both these process nodes from consideration. Thus, zonalists accepted parent material as one of the five factors of soil formation, which was derived from bedrock by epimorphism. In other words epimorphism was recognized, but only as a prepedological process. It was the same with near-surface processes of erosion and deposition, for they were recognized as geomorphological and excluded from pedological consideration (Nikiforoff 1949, Butler 1959). However, far from being able to ignore these two process nodes, Part I established that they exercised a crucial control on the nature of the resulting soil material. In addition, in order to understand the global distribution of soil material, it is necessary to relate these same process nodes to the factors of soil formation. As stated in the Introduction, five such factors are proposed: parent material, topography, climate, the biosphere and time. But they had been defined in a zonalistic way and before attempting the suggested correlation it is necessary to reassess them in terms of this very much broader process-orientated model now under consideration. From what has been said above, parent material cannot operate as a factor in these circumstances and it is proposed to replace it with **lithospheric material**. Climate also needs to be reassessed, for in terms of the two process nodes it is too vague to achieve any worthwhile correlations. In these terms the only climatic parameter that is essential is the presence of water, for without it there is no possibility of weathering, leaching or new mineral formation taking place, or of rainwash being operative. Therefore, it is proposed to replace climate as a factorial descriptor with **availability of water**. The other factors of topography, the biosphere and time do not require amendment.

It is now possible to consider the relationship that exists between each of these factors – lithospheric material, topography, availability of water, biosphere and time – and the epimorphic and near-surface process nodes. It would be best to start from a consideration of time, and what is of fundamental concern in this case is the rate of formation of soil material. Although this is not known with any degree of preciseness, it is apparent that the two process nodes operate within different time frames. From the results given in Chapter 3 it would seem that topsoil of some 20 cm in thickness can be formed as a result of the operation of near-surface processes over some 10^2–10^3 years. However, estimates of the rate of saprolite formation (Saunders & Young 1983, Young & Saunders 1986, Pavich 1989) indicate that a comparable thickness would require 10^4–10^5 years to form. In other words there is at least an order of magnitude difference between the rate at which topsoil and saprolitic subsoil are formed, so that in considering soil formation it is essential to separate the impact of near-surface and epimorphic processes.

In addition the larger figure can be taken as being the average overall time required for soil formation. That being so, it is possible to use it as a standard by which to consider the temporal variability of the other factors of soil formation. Thus, in most environments the nature of lithospheric materials and topography would not vary to any extent over such a period of time and hence both of them must have a strong influence on the nature of any soil material that develops. Therefore, it is possible to regard both of these as **determinative factors** in soil formation. There is of course a major difference between them, for lithospheric materials are mainly involved in epimorphic processes and the production of saprolite, whereas topography is more concerned with the control of near-surface processes involved in the formation of the topsoil.

In contrast to these determinative factors, the availability of water and the nature of the biosphere, over a comparable time interval (10^4–10^5 years), have been subject to a much greater degree of variation. Thus, Figure 7.1 is an estimate of the change in the extent of the African rainforest between the present and 18 000 years ago (Butzer 1978). Given such a high rate of change, it is impossible to ascribe to these factors too much determinative influence on the resulting soil materials. However, without their presence none of the processes, either epimorphic or near-surface, would be able to operate. It is possible then to think of these factors as operating an on/off

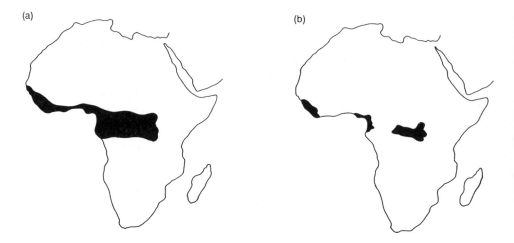

Figure 7.1 The African rainforest (after Butzer 1978): (a) present day; (b) 18000 years ago (inferred).

pedogenic switch and to regard them as **initiating factors** in soil formation.

Of course factors of soil formation do not act independently but always operate together as a complex. In its simplest form this may be expressed as follows: lithospheric materials at a particular site (the determinative factors) require the presence of water and the biosphere (the initiating factors) so that epimorphism and near-surface processes can operate over time to produce soil material. In other words the formation of soil material results from the operation of epimorphism and near-surface processes within boundaries determined by the interacting factorial complex. From this it is apparent that soil materials, processes and factors form an ascending hierarchic sequence, within which processes have a pivotal position (Fig. 7.2), for it is impossible to move directly from factors to materials, as was attempted in the zonal scheme, without considering processes in the expanded form.

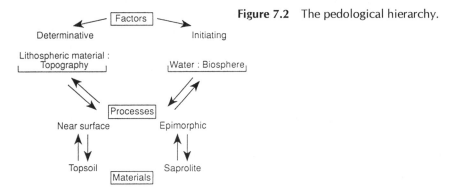

Figure 7.2 The pedological hierarchy.

110

It is apparent from this discussion that the primary factors controlling the global distribution of soil are the determinative ones of lithospheric material and topography. In turn the nature of these determinative factors is controlled by plate tectonics (Paton 1986). This concept recognizes that the surface of the solid Earth is composed of interlocking rigid lithospheric plates, which float in an underlying plastic asthenosphere. The plates are in slow but constant motion, moving away from one margin, the mid-oceanic ridge, where the lithospheric material is formed from eruptive basalts, towards the opposite margin, where it is resorbed into the mantle along a subduction zone. Thus, it is possible to consider any continental lithospheric plate as being made up of three segments, a tensional margin, a plate centre and a compressional margin, within each of which the determinative factors of soil formation are very different (Table 7.1).

Table 7.1 Plate segments and environments.

	Plate segments		
Environmental influence	Tensional margin	Plate centre	Compressional margin
Lithospheric material	Basaltic	Granite	Mixed
Topography	Steep with plateaux	Gentle	Steep
Volcanism	Active non-explosive	None	Explosive
Seismicity	Localized	Weak	Strong regional

Plate centres cover by far the greatest area and within them lithospheric material is overwhelmingly granite, or granite-derived, and topography is gentle. It also needs to be emphasized that for at least the past 250 Myr, since the beginning of the Mesozoic, continental plate centres have maintained themselves as land areas exposed to pedogenic processes, except for interruptions by shallow marine transgressions in the mid-Cretaceous and the more recent Pleistocene glaciations. At the beginning of the Mesozoic the continents were assembled in one landmass, Pangaea, or two closely associated supercontinents of Laurasia in the northern hemisphere and Gondwana in the south (Fig. 7.3), which have since fragmented and dispersed leaving the Atlantic and Indian Oceans in their wake (Fig. 7.4). In the process the continents have moved great distances and been subject to a great range of conditions. This means that pedogenic processes have been operating over immense periods of time and, even though this must have been somewhat sporadic in view of the variable input of the initiating factors, it must mean that the processes of soil formation could have reached their endpoint many times over. This has led to the accumulation on each of the continental fragments of considerable amounts of inert endproducts, such as fine-grain kaolins, iron and aluminium oxides and most particularly quartz sands, so that

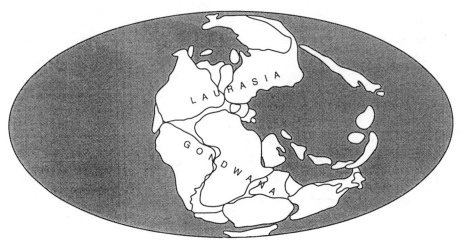

Figure 7.3 Pangaea, 250 Myr ago.

present-day pedogenesis on continental plate centres is for the most part secondary rather than primary and inheritance is dominant.

Compressive continental plate margins form the circum-Pacific, Asian and Mediterranean mountain belts, which are peripheral to the former supercontinents of Laurasia and Gondwana (Fig. 7.4). Their lithospheric materials are of very mixed composition. They include poorly sorted and weakly epimorphosed greywackes and metamorphosed bedrocks, all of which have been shattered by Earth movements. Volcanic ash, which can range in composition from rhyolitic to basaltic, is also important. Such a variety is very different from the granitic uniformity of the plate centres. This contrast is further emphasized by the very steep topography, for this results in the dominance of mass movement, which occurs at such a rate that there is insufficient time for any great degree of pedogenesis to occur, even though the nature of the lithospheric materials makes them very susceptible to epimorphism.

Tensional margins are of little pedological importance, for they are generally submerged beneath the deep ocean. In the few circumstances where they do emerge above the ocean surface, the basaltic nature of the lithospheric material is the dominant factor in determining the nature of the resulting soil material.

In view of the overwhelming importance of soil development associated with continental plate centres they will be dealt with in some detail in the next three chapters (Ch. 8 on Australia, Ch. 9 on Africa, and Ch. 10 on other Gondwana and Laurasian centres) before considering the soil developed at continental plate margins in Chapter 11.

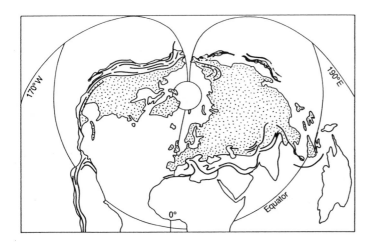

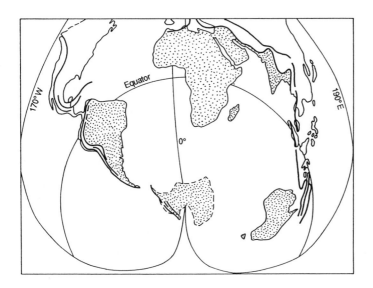

plate centres   compressional plate margins

Figure 7.4   The present-day consequences of Pangaean dispersal
(after Holmes 1965).

CHAPTER 8

Soils of a continental plate centre, I: Australia

Conceptual locks are far more powerful than factual locks as barriers to scientific understanding. *Stephen J. Gould*

The whole of the Australian continent is on a plate centre, for its tensional or ridge margin lies far to the south in the Southern Ocean and its compressive margin a long way to the north and east in New Guinea and New Zealand. Being a continental plate centre the bedrock of Australia is for the most part granite or granite-derived (i.e. sedimentary rocks such as sandstones, shales and low-grade metamorphic equivalents), its topography is gentle and in the more recent past volcanism and seismic activity have been minimal. Given its long-term stability and exposure as a landmass, there has been sufficient time for the full development of the effects of epimorphism and near-surface processes, so that texture-contrast soils should be a very common product of pedogenesis. However, this was not immediately apparent to Australian pedologists, for the fundamental significance of the abrupt texture break between the topsoil and subsoil was not realized. Soils that are now recognized as texture-contrast were classified on the basis of a variety of other criteria, so that they were scattered over at least 11 great soil groups (Stace et al. 1968; Table 8.1). The idea of a soil map of Australia, based on

Table 8.1 Great soil groups with texture contrast profiles (derived from Stace et al. 1968).

Harsh subsoils	Mellow subsoils
Solonetz	Grey-brown podzolic
Solodized solonetz	Red podzolic
Soloths	Yellow podzolic
Solonized brown soils	Brown podzolic
Red-brown earths	Lateritic podzolic
	Gleyed podzolic

Northcote's *Factual key* (1960), began to change this view, for it brought the concept of a sharp texture contrast to the fore. With the mapping for the *Atlas of Australian soils* (Isbell et al. 1967) it became apparent that there was a particular association of these texture-contrast soils with middle and lower hillslopes and valley margins throughout large areas of Australia (Thompson & Paton 1980). This association indicated a common mode of origin related to landscape position, and was one of the factors that led to the evaluation of the pedological processes discussed in Part I of this book.

Within the group of texture-contrast soils there is a great deal of morphological variation, but much of it can be rationalized in terms of whether the clayey subsoil is mellow or harsh (Appendix 3), for such a difference also has an effect on the topsoil. Thus, along the topsoil/subsoil interface, a harsh type of subsoil disperses with relative ease on coming into contact with water, with the physil particles so detached being pervected (Paton 1978) and deposited along subsoil planar voids, restricting further through-drainage. Such impedance is rapidly induced and may occur repeatedly over a short time period, giving rise to an intense and sharply demarcated bleach at the topsoil/subsoil interface. In comparison mellow subsoils are highly stable when in contact with water and hence are subject to little if any dispersion. Impedance in this case occurs only in texture-contrast soils in lower topographic situations affected by the slow rise and fall of the seasonal water table, or during episodic periods of heavy rainfall. This results in a much less intense topsoil bleaching, spread over a broader zone with less sharply demarcated boundaries. Soil moisture draining to the topsoil/subsoil interface is constrained to flow laterally along the boundary, and hence any materials carried in solution or suspension are also carried along this interface. Such flow is generally slow and only fine-grain mineral particles are moved in suspension. With periodic evaporation anything in solution or suspension is deposited, generally in the subsoil planar voids just below the interface, giving rise to a range of subsoils, which can be rich in carbonates, iron, silica or physils.

In addition to their widespread development on lower and middle hillslopes, a great number of texture-contrast soils occur in areas of valley fill. This is to be expected, for often there is little if any break of slope between the valley fill and adjacent hillslopes, so there is nothing to prevent the mobile topsoil transgressing this boundary. Figure 8.1 shows a typical example from the Bremer Valley, 50 km west of Brisbane in southern Queensland, where low-angle hillslopes with texture-contrast soils merge almost imperceptibly into broad areas of valley fill. In this case the topsoil from the hillslopes continues across the upper (third) terrace, where it transgresses a variety of depositional clays (Paton 1965). In places its continuity

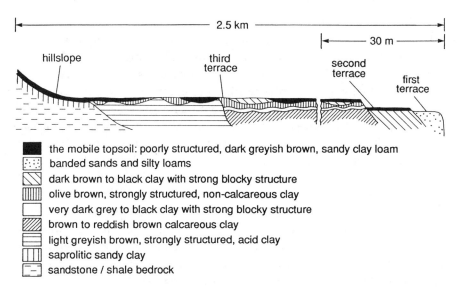

the mobile topsoil: poorly structured, dark greyish brown, sandy clay loam
banded sands and silty loams
dark brown to black clay with strong blocky structure
olive brown, strongly structured, non-calcareous clay
very dark grey to black clay with strong blocky structure
brown to reddish brown calcareous clay
light greyish brown, strongly structured, acid clay
saprolitic sandy clay
sandstone / shale bedrock

Figure 8.1 Texture-contrast soils and river terraces (after Paton 1965).

is interrupted by mukkara where fine-grain sediments of the valley fill penetrate upwards through the more massive topsoil. It is also possible to trace
this transgressive topsoil down onto the second terrace before it is disrupted
by the effects of alluvial cut and fill. At the same time it is possible in some
cases to recognize a facies change in the topsoil. In this case the finer-grain
materials, brought to the surface by bioturbation and mukkara formation,make a greater contribution to the topsoil, so that in places the texture
contrast is replaced by a fabric-contrast soil. The balance is different in
areas where, despite texture-contrast soils being formed on hillslopes,
stream activity is sufficient to truncate the distal end of the mobile topsoil
at the end of the footslope so that the nature of the valley fill is entirely determined by alluvial processes. Such situations are known, for example, in the
Canberra area (Hesse 1985), and occur sporadically throughout eastern
Australia especially in steeper low-order catchments.

It is also possible, as was seen in Chapter 5, to extend the formation of
texture-contrast soils into the more arid areas of the Australian continent
where the effects of rainwash, bioturbation and vegetation cover retain their
potency to enforce the formation of texture-contrast soils. Even in those
situations where there is insufficient water to maintain pedogenesis, the general effect will be one where the results of prior pedogenesis are preserved
without a great deal of subsequent alteration, for wind is the only agent that
can remain active, and as seen previously (Ch. 5) the energy available to it
is strictly limited. In general, the role of wind erosion in these soils will be

117

limited to the removal of the topsoil, down to any stonelayers. The residual quartz sands would be winnowed of fines and the coarser material accumulated as mobile sand sheets and dunes capable of burying the previous landscape and associated soil materials. The great areas of central Australia covered by **siliceous** and **earthy sands** represent the accumulative endpoint of granite pedogenesis, which has been continuously in operation for at least 250 Myr since the Permo-Triassic when Australia formed part of Pangaea.

Given a predilection for splitting by taxonomic pedologists, based on morphological differences of unknown relevance (Table 8.1), this group of texture-contrast soils has given rise to a plethora of terms, which has caused a great deal of confusion, particularly as many of the names are not exclusive. Rationalization can be achieved by first of all recognizing the fundamental texture contrast, then the nature of the saprolite and whether anything has been added to it, before considering the biomantle and its degree of bleaching. In other words a progressive analysis, based on process, will help to abolish the tyranny of unnecessary names.

Discussion so far has concentrated on soils where the topsoil is often not more than 40–50 cm thick, so that within a normal type of soil pit it is possible to see topsoil, saprolite and, depending on availability of suitable clasts, a stonelayer at the base of the former. In fact for many researchers it is only in this context that a texture contrast is recognized. Yet topsoil thickness can vary greatly, for all that has to happen is a change in the rate of its formation. Thus, on Cape York Peninsula (Isbell & Gillman 1973; Fig. 8.2) deep sandy soils have accumulated on very gentle surfaces of low-level granite plateaux as a result of epimorphism, mesofaunal activity and near-surface winnowing of fines, but with very little if any lateral movement of the topsoil as a whole. The soil consists of medium sand with a very weak fabric development down to almost 1 m, below which there is a boundary grading into a coarser sand with rounded quartz gravel 2–4 cm in diameter, which is an ill defined stonelayer. At a depth that varies from 1.3 to 2.0 m, there is a clear boundary to a saprolitic gritty grey clay, with coarse red mottles, which is generally at least 50 cm thick before it grades into granite bedrock. In general pedologists have regarded this as a deep sand because they rarely investigated beyond 1 m depth. However, by going down to bedrock granite, the soil is clearly a texture contrast with a thick biomantle, which has accumulated in an area of gentle topography.

The build-up of a mobile topsoil can become very marked when suitable quartz-rich bedrock occurs in a landscape of very gentle relief and where precipitation is not all that great. These conditions occur on the gently undulating plains of the seasonal tropics such as in the Torrens Creek area of north Queensland, some 200–300 km southwest of Townsville (Coventry

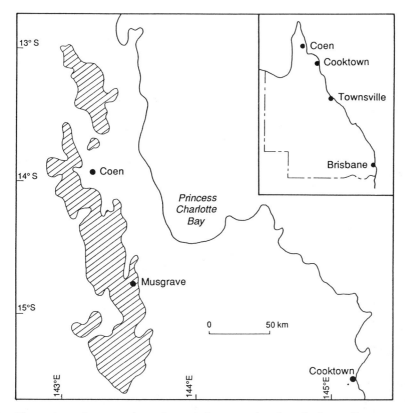

Figure 8.2   Deep sands on Cape York Peninsula (after Isbell & Gillman 1973).

1982; Fig. 8.3). The bedrock consists of gently dipping Permian and Trias-
sic sedimentary rocks and on these texture-contrast soils have been formed
in the normal way, as described previously. Because of the gentle slopes and
low precipitation there is minimal movement of topsoil down slope. However, winnowing occurs to remove the fines, leaving the coarser sand particles to accumulate and form **yellow, grey** or **red earthy sand** topsoils, of
1–3 m in thickness (Fig. 8.4). In places these grade laterally into texture-
contrast soils with topsoils of a more normal thickness (≈50 cm). The winnowed fines are deposited in the broad open drainage lines where they accumulate to depths in excess of 20 m in places, for there is no way of removing
them farther down stream under existing streamflow. The fine-grain nature
of the material enables a massive, **earthy red loam** to develop, in which
there is a lack of obvious sedimentary features. This suggests slow accumulation because of the gentle relief, where any original sedimentary features
would have been destroyed by faunal activity. Overall, however, the most
important point to note is that, despite a considerable change in environ-

119

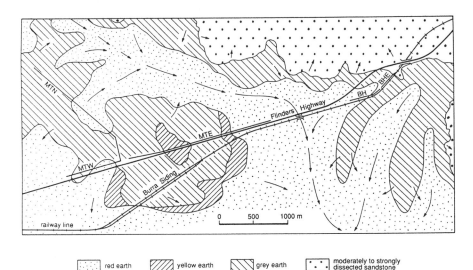

Figure 8.3 Map of deep accumulative texture-contrast soils, north Queensland (after Coventry 1982).

mental conditions from the humid zone of the Great Dividing Range to these semi-arid conditions, the fundamental pedogenic processes remain much the same: the formation of texture-contrast soils.

Another example of a texture-contrast soil with a thick topsoil (Fig. 8.5) comes from the Alice Tableland on the continental divide some 200 km south of Torrens Creek. Here Gunn (1967) describes the occurrence of **red** and **yellow earths** several metres deep with ironstone nodules concentrated towards the sharp lower boundary where they overlie a saprolite-derived from quartzose sandstones and siltstones, which is indurated by iron oxides and hydroxides where it outcrops. In other words these soils are the same in form and genesis as the soils of Torrens Creek. Such soils are common throughout eastern Australia and their occurrence extends westward into the Northern Territory.

Before leaving this example from the Alice Tableland, it should be noted that the region provides an epitome of texture-contrast soils formation, for not only are there deep accumulative texture-contrast soils on the tableland of the continental divide (Fig. 8.5, zone 1), but also in the zone immediately east of the tableland more normal texture-contrast soils occur (Fig. 8.5, zone 3). Dissection at this point has cut through the deep accumulative texture-contrast soil and a sequence of Tertiary and possibly Cretaceous sediments to expose a variety of subsoils, which are now covered by a relatively thin topsoil that varies from being coarser-textured (sands) up slope and finer-textured (loams) down slope, so that the sequence can be interpreted as consisting of a mobile topsoil moving down slope across the varied

120

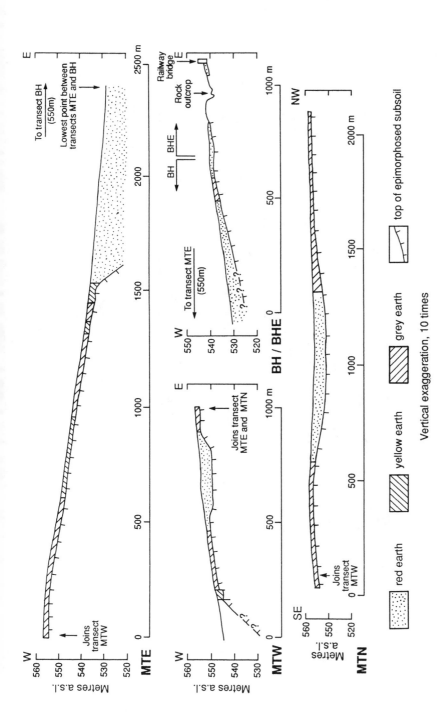

Figure 8.4 Sections of deep accumulative texture-contrast soils in northern Queensland (after Coventry 1982).

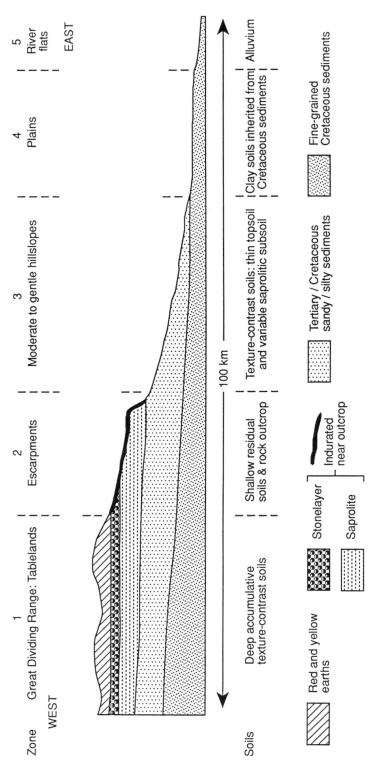

Figure 8.5 Great Dividing Range, northern Queensland, and texture-contrast soils (after Gunn 1967).

subsoil and being sorted in the process to form texture-contrast soils of the kind discussed earlier in the chapter.

A comparable situation also occurs in the southwestern portion of the continent on the Precambrian shield areas of Western Australia. In this case the topography varies by no more than 60 m, over distances of 40–50 km, between the smooth rounded uplands and the shallow, wide, ill defined valleys. The uplands are the dominant part of the landscape and are largely featureless, except for some granite bosses found occasionally on the highest parts of the divides, or more commonly on the valley slopes (Mulcahy 1967). These uplands are generally referred to as sand plains, for they are covered to a depth of several metres by a **yellow earthy sand**, which for the most part is uniform except for a concentration of coarser sand and iron-stone fragments towards its base. Across a sharp boundary this surface sand overlies saprolitic, granitic bedrock, with the depth to fresh granite being very variable. Fabrics are more sandy, because they are derived from granite, compared to the more earthy fabrics in north Queensland where the bedrock is generally finer-grain sedimentary rock. It is possible to recognize this as a texture-contrast soil where the yellow sand forms a very thick topsoil, with a stonelayer characteristically developed near its base. The underlying granitic saprolite forms the subsoil. As a result of the concentration of water flow along the topsoil/subsoil interface, a considerable amount of iron has been transported and is eventually precipitated as oxides and hydroxides to form an indurated zone in the subsoil immediately below the interface where this nears the surface. The landscape as a whole can then be interpreted as being one that is being swamped by the accumulation of the residual quartz sand fraction from the granitic saprolitic subsoil as a result of near-surface winnowing, in much the same way as in the previous examples from north Queensland. The interpretation that contemporary processes are affecting the topsoil receives support from the work of Brewer & Bettany (1973), who showed that the skeleton grains of the soil consisted not only of sand-size quartz grains as expected, but also of rather brittle sand-size kaolinitic glaebules (Brewer 1964), which occurred as discrete entities randomly packed with the quartz grains. The presence of these glaebules pointed to the topsoil being relatively young, and indeed Brewer & Bettany concluded that this yellow earthy sand had been formed in a sedimentary environment in which movement was restricted, probably as a colluvial slope deposit. Furthermore the uniformity of distribution of the glaebules throughout the yellow earthy sand, which showed little if any breakdown, suggested to them that soil formation was at an early stage and that the soil could be explained in terms of a first-cycle derivation from the underlying saprolite. In terms of the model set out in Part I, a simpler

123

explanation is that the glaebules have been derived from the subsoil by contemporary mesofaunal activity and there is no need for recourse to various cycles of formation. This point needs to be emphasized, for these soils are generally regarded as "fossil" and yet the topsoil is still being subjected to pedogenesis. The same is also true of the red and yellow earths of the Alice Tableland, for even where they are being strongly dissected, contemporary pedogenesis continues on the tableland surface.

It is apparent from this discussion that shallow and deep texture-contrast soils together with residual quartz sands are very widespread on the Australian continent. Outside of this only the clay soils need to be accounted for, of which there are three distinct groups. In the first of these the clays are derived from basaltic lavas, erupted by a series of Tertiary and Quaternary volcanoes along the Great Dividing Range near the east and southeast coasts. The basic lavas of the volcanoes are responsible in turn for producing two kinds of clay soils. In the first the minerals are affected by strong weathering and leaching and the epimorphic products consist mainly of kaolins together with oxides and hydroxides of iron and aluminium, strongly coloured in reds, yellows and browns, which combine to form extremely fine and very strong peds, which are markedly subplastic (see Appendix 3). The second type of clay soil associated with basalt is very different, for in this case the minerals have been altered by hot late-stage solutions into a mixture of smectite clays and carbonates. The clay soil derived from these materials involves direct inheritance, with little sign of any pedogenic process, beyond a possible movement of carbonate to the subsoil and down slope (see Fig. 4.13). The resulting soil is dark-coloured, smectite-rich, with coarse dense peds, which is highly plastic, expands and contracts on wetting and drying and in places develops mukkara structures.

In the second group the clays are directly inherited from fine-grain Cretaceous sediments, which have resulted from the only major marine transgression to which Australia has been subject since it formed part of Gondwana. The soils are smectite-rich, coarsely structured, grey and brown clays frequently associated with carbonate, directly inheriting most of their characters from the bedrock. Such clays are restricted of course to those areas where the Cretaceous sediments have been recently exposed by stripping of the overlying soils and sediments. This can be seen in Figure 8.5 (zone 4) where dissection has cut down through the slightly lithified deposits making up the Alice Tableland, until the Cretaceous clays have been reached. However, this is only a relatively small area east of the Great Dividing Range. The greatest expanse of these clay soils occurs to the west of the divide, along a broad arc from northern New South Wales through midwestern Queensland and into the Northern Territory.

124

The third group of clay soils are those that result from the deposition of fines that have been removed by rainwash, as suspended load, during the formation of texture-contrast soils. They have been deposited along the floodplains of streams that drain areas where texture-contrast soils are developed and are particularly common in distal parts of riverine plains such as the Riverina and the Gulf of Carpentaria.

The uniform texture of these clay soils precludes the development of texture-contrast features, but as seen in Chapter 4 a mobile topsoil with a different fabric from the subsoil is common and is responsible for the development of fabric-contrast soils (see Figs 4.12 and 4.13; Plate 5).

Before closing this chapter, consideration should be given to situations where accumulations of a considerable thickness of quartz-rich, inert, mobile topsoil have been responsible for the formation of **podzols**. In normal circumstances the highly active organic molecules produced by the breakdown of plant and animal materials are inactivated by clays to produce the clay–humus complex so characteristic of the surface of many soils, but this is not possible where deep pure quartz sands have accumulated. The only materials available for the organic molecules to react with are the fine-grain iron oxides and hydroxides, which very commonly form a skin around individual quartz sand grains. The organic molecules form a chelation compound with the iron, which as a result is removed from the quartz grains, giving them a bleached appearance. The iron is transported in the chelate form deeper into the quartz sand body where it is redeposited as a pan, so that a podzol is produced. Bloomfield (1953, 1954) indicated that the active organic molecules were simple polyphenols produced by a whole range of plant materials, and Handley (1954) showed that the simple polyphenol content in plants was related to the nutrient status of the soil material: the poorer and more acidic the material (i.e. the more quartz-rich), the higher the content of simple polyphenols. Coulson et al. (1960) and Davies et al. (1964) showed that the maximum quantity of simple polyphenols occurred in freshly growing green leaves, with much less in senescent or dead leaves and minimum amounts in the humus layer on the soil surface. Thus, it can be appreciated that the production of podzols is dependent upon special circumstances (see Plate 6).

Deep accumulations of quartz sand occur in restricted sites on the surface of Hawkesbury Sandstone plateaux in the Sydney region as a result of the downslope movement of mobile sandy topsoils combined with the inability of rainwash and streamflow to remove the accumulations. It is within such accumulations that podzols have developed (Buchanan & Humphreys 1980, Buchanan 1980). On the Lambert Peninsula, which projects into Broken Bay (Fig. 3.1), 28 such sites of podzol development have been located, and Figure 8.6 shows a typical example where a thick sand deposit has accumulated at the base of a hillslope. This typical two-pan podzol is about 2 ha in extent and has developed on either side of the stream coincident with the thickest sand deposit, which is well drained as a result of the stream having incised into it. The surrounding non-podzolized yellow-brown sand is both thinner and less well drained. The podzol

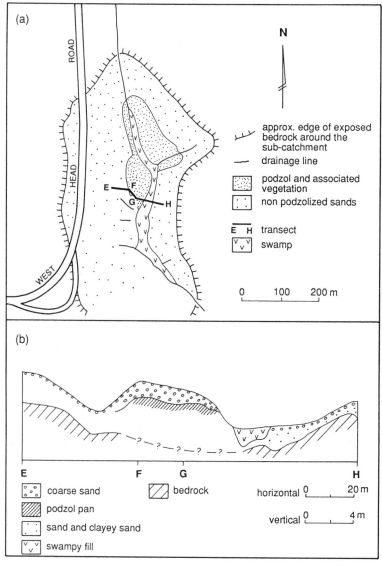

Figure 8.6 Deep sand topsoils and podzols (after Buchanan & Humphreys 1980).

supports a woodland of *Eucalyptus gummifera, E. piperita, Angophora costata* with *Ceratopetalum gummiferum* as a conspicuous understorey species. In contrast the surrounding non-podzolized sand supports a low woodland mostly dominated by *E. haemastoma, E. oblonga, E. gummifera* and *E. sieberi*.

Suitable conditions for podzol formation also exist within coastal dune-fields but decalcification of the sand body is required before podzol development becomes well established. As very thick accumulations of quartz sand can occur in this setting the podzol can be exceptionally deep with a bleached zone extending over many metres. These so-called giant podzols occur from place to place on the east Australian coast. It needs to be stressed that, despite the fact that podzols have what is conventionally regarded as a well developed, mature, profile, they can form in a very short time (Paton et al. 1976). This is because the only process involved is a simple stripping of the very thin covering of iron oxides and hydroxides around the quartz sand grains by chelation and the downward transport of a relatively small amount of material to form the pans.

This chapter has attempted to show that the greater part of the soils of the Australian continent are explicable in terms of their derivation from a granitic bedrock, which has been affected by epimorphism and near-surface processes, under conditions of minimal topographic variation to give rise to a texture-contrast soil with either a thin or a thick topsoil (Fig. 8.7). Even though clay soils are morphologically very different they have been subject to the same set of processes, but in this case leading to fabric-contrast rather than texture-contrast soils.

To establish similarities between Australia and other continents it is sufficient, therefore, to demonstrate that the granitic strand of pedogenesis can be recognized. This is attempted in the next chapter with respect to Africa.

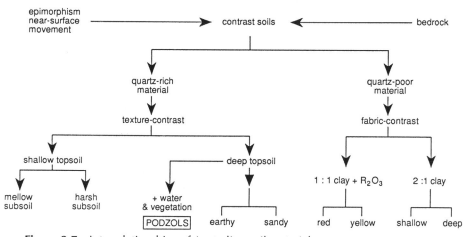

Figure 8.7 Interrelationships of Australian soil materials.

127

CHAPTER 9

Soils of a continental plate centre, II: Africa

One cannot arrive at explanations without using one's own personal judgement and this is inevitably subjective. A subjective treatment is usually far more stimulating than a coldly objective one because it has a greater heuristic value. *Ernst Mayr*

Africa is another Gondwana fragment, another plate centre, where granite and granite-derived materials are the dominant bedrock, topography is for the most part gentle, whereas volcanism and seismicity are at a minimum, apart from the East African Rift and the Cameroon area. These broad environmental conditions are very similar to those of Australia, and therefore soil materials can be expected to have common properties. This similarity is seen at its simplest and most straightforward in the eastern Sudan, near the Ethiopian border, where steep-sided granite hills, with fringing sandy pediments, rise 50–150 m above extensive clay plains (Ruxton 1958). Vegetation on the plains is of the acacia/tall-grass type and rainfall, essentially summer thunderstorms, amounts to some $650\,\mathrm{mm\,yr^{-1}}$.

A section of almost 300 m was examined from the top of one of these granite hills across the upper pediment and onto the lower hillslope (Fig. 9.1). Three layers of material were distinguished overlying the granite bedrock (Tables 9.1, 9.2) according to their quartz:feldspar ratio, which was

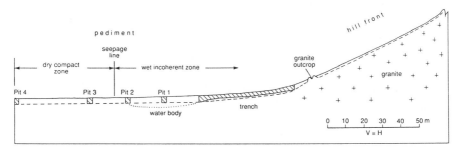

Figure 9.1 Sudan, Ruxton's catena (after Ruxton 1958).

129

Table 9.1 Sudan, trench/particle-size analyses and mineralogy (after Ruxton 1958).

Sample[1] No.		Gravel (%)	Sand (%)	Largest grain (mm)	Quartz (%)	Quartz: feldspar[2] ratio
Topsoil (layer 1)						
	Up slope					
8	↓	44	33	20	39	1.0
7	↓	59	24	30	45	1.2
6	↓	62	28	30	51	1.3
5	↓	36	47	14	47	1.3
4	↓	35	46	9	56	2.1
3	↓	41	41	15	61	2.8
2	↓	27	57	8	55	1.8
1	Downslope	24	55	7	57	2.4
Upper subsoils (layer 2)						
	Up slope					
16b	↓	62	18	14	67	4.8
16a	↓	73	13	20	61	2.3
15	↓	53	21	10	65	6.5
14	↓	60	17	10	69	7.7
13	↓	74	11	9	78	9.8
12	↓	60	18	12	71	10.0
11	↓	55	20	9	63	4.9
10	Down slope	58	20	9	70	7.8
Lower subsoil (layer 3)						
	Up slope					
25	↓	64	21	25	28	0.5
24	↓	59	26	15	44	1.0
23	↓	63	23	18	28	0.5
22	↓	44	32	12	43	1.3
20	↓	50	25	10	51	2.0
19	Downslope	51	24	15	27	0.6

1 Sample number as given in Ruxton (1958, table 1).
2 The quartz:feldspar ratio determined from all material coarser than 200 mesh sieve (0.076 mm aperture).

low in layer 1 (the topsoil), high in layer 2 (the upper subsoil) and low again in layer 3 (the lower subsoil). The source of the fresh feldspar in the topsoil was not the lower subsoil, for the low feldspar zone of the upper subsoil always intervened; it must have been derived by lateral accession from the granite outcrop up slope, where it was observed that boulders broke down into compound quartz/feldspar, gravel-size fragments and then into individual mineral grains as they moved down slope. In other words, the topsoil was mobile across the underlying materials, which had been subject to *in situ* epimorphism, the most extreme products of which occurred in the upper subsoil, or zone I saprolite, whereas the lower subsoil, or zone II saprolite, showed less extensive effects of epimorphism as the bedrock was approached. The downslope movement of the topsoil was also reflected in the

Table 9.2 Sudan pits particle-size analyses and mineralogy (after Ruxton 1958).

No.	Layer (cm)	Gravel (%)	Sand (%)	Largest grain (mm)	Quartz (%)	Quartz: feldspar ratio
Pit at downslope end of trench						
34	1 (0)	27	52	8	49	1.5
33	2 (60	65	16	8	71	6.5
32	3 (150)	71	14	10	80	16.0
31	3 (225)	70	50	10	75	4.7
30	3 (315)	58	25	18	27	0.5
Pit 1						
39	1 (0)	19	49	6	53	2.9
38	1 (54)	56	21	10	64	4.6
37	2 (105)	66	15	8	80	32.0
36	3 (180)	65	16	10	73	7.3
35	3 (255)	63	20	10	57	2.2
Pit 2						
42	1 (0)	26	51	7	60	3.0
41	2 (75)	68	12	10	80	160.0
40	2 (165)	83	14	11	96	192.0
Pit 3						
46	1 (15)	26	51	5	62	3.7
45	2 (90)	70	11	10	81	81.0
44	2 (180)	65	14	10	68	7.6
43	3 (270)	72	18	30	28	0.4
Pit 4						
51	1 (0)	19	53	5	56	3.1
50	1 (45)	46	31	7	69	7.6
49	2 (99)	64	13	10	76	38.0
48	2 (189)	61	15	8	71	11.8
47	3 (283)	55	21	11	61	4.1

surface concentration of feldspar, the inverse downslope relationship between feldspar and clay, and the gradual concentration of the gravel component towards the base of this layer so that a much more clearly differentiated stonelayer was developed farther down slope. All of these factors gave rise to a marked discontinuity between the topsoil and the saprolitic subsoil, which was also evident when feldspar amounts were plotted against depth across the lower pediment (Fig. 9.2). There was a great deal of water movement along the interface between the topsoil and the subsoil, as a result of thunderstorm rain draining off the hillslope, which was responsible for considerable lateral pervection of the upper subsoil, or zone I saprolite, down slope along the interface, until it emerged as seepage on the lower pediment. These pervected fines were deposited as a thin, patchy, dark clayey silt, which consisted of 21% sand, 58% silt and 21% clay, with a quartz:feldspar

131

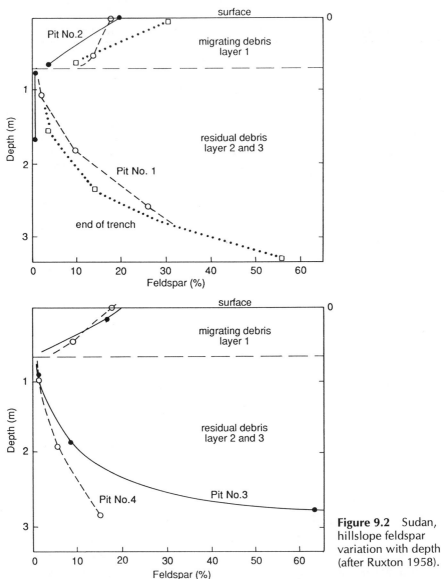

Figure 9.2 Sudan, hillslope feldspar variation with depth (after Ruxton 1958).

ratio of 20, which showed that it had been derived from the upper subsoil, or zone I saprolite.

Thus, in the eastern Sudan a sequence of soils is developed in which a coarse-textured, laterally mobile topsoil, with a basal stonelayer in lower topographic situations, abruptly overlies a finer-textured, granitic, saprolitic subsoil. This is almost identical, both in form and in the processes invoked, to that discussed in the case of Killonbutta in central New South Wales (Fig. 4.11). In other words a texture-contrast soil has developed in the Sudan by exactly the same processes invoked in central New South Wales.

The correlation between process and the resulting soil materials in the Sudan and central New South Wales was possible because in both cases stratigraphic methods were used to achieve lateral correlation of soil layers across a landscape. However, in the rest of Africa such a correlation was rather more difficult, for pedological thought was dominated to a much greater extent by zonalistic ideas with its concentration on processes operating vertically downwards from the surface to produce profiles. Specifically excluded from consideration were the lateral surface processes of erosion and deposition that were so central to both the Sudan and Australian examples, for, according to zonalists, such processes were geomorphological and thus were excluded from pedological consideration (Nikiforoff 1949, Butler 1959; see Introduction). However, as a result of the work of Geoffrey Milne in East Africa (Milne 1935, 1936, 1947) this obsession with the vertical dimension and the rigid demarcation between pedology and geomorphology was weakened. The most important of Milne's innovative concepts was the **catena** (Milne 1935), which referred to a repetitive soil pattern found in East Africa on uniform, granitic bedrock, extending from red loams on hillslopes to black clays in intervening depressions. Initially the main use of the catena was as a complex mapping unit, but much more fundamentally it made pedologists look at profile differences within a landscape. Even though to begin with only differences of drainage were considered, it moved the centre of interest from zonalistic verticality to lateral downslope variation. The idea was subsequently expanded (Milne 1936, 1947) to include normal surface erosion as a pedological process in catena development, which necessitated discussion of soil variation in terms of a continuous section running from drainage divide to streamline, rather than descriptions of isolated profiles. Milne gave as an example one that was widespread throughout central Tanzania (Fig. 9.3). It consisted of grey loams around granite outcrops on the drainage divide, below which occurred a sequence of soils going from red up slope to yellow down slope, with a parallel change in texture grade from sandy clay and sandy clay loam to loamy sand and sand. Milne recognized that these changes were the result of the operation of surface processes of erosion, transportation, sorting and deposition. Within the upslope, heavier-textured, red soils, there was a complete range of particles that the granite could yield as a result of epimorphism, from quartz sand to feldspar-derived clay. This material was eroded by rainwash and the particles differentially transported down slope. The coarser quartz particles were retained on the hillslope as a yellow-grey sand and the finer particles were transported out of the system, or deposited in the drainage lines. Particular emphasis was given to the fact that erosion, transport and deposition were slow and non-catastrophic, so that the material behaved

133

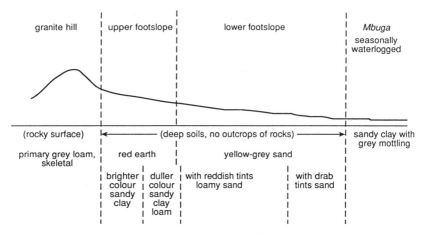

Figure 9.3   Tanzania, Milne's catena (after Milne 1947).

throughout as a soil and at no time would a non-pedological colluvium be formed. In other words the importance of contemporary erosion/deposition on hillslopes as a pedological process was recognized by Milne. In addition he also recognized the long-term consequences of the retention of coarser sand particles on lower hillslopes, for he referred to the endpoint of such sand accumulation, in southern Tanzania, as burying the landscape and forming gently undulating sand plains, with occasional projecting bosses of fresh granite. This description is very like the sand plains of southwestern Australia, referred to in Chapter 8.

The recognition of hillslope erosion as an integral part of pedogenesis immediately created a paradox, for the original catena concept dealt with differences in drainage conditions within profiles, which could be melded into the zonal, profile-restricted model with ease. However, the processes of hillslope erosion, involving the movement of solid material from one profile site to another, could not, and unfortunately this matter was not resolved before Milne's premature death in 1942. Despite this, however, Milne's influence remained very considerable among British pedologists working in Central and East Africa where soils similar to those he recognized in Tanzania were identified as **plateau soils** over great areas of Zambia (Trapnell & Clothier 1937, Trapnell 1943) and Zimbabwe, which subsequently gave rise to three detailed pedological investigations that illustrate very well the equivocation resulting from this paradox.

A catena of soils on granitic, Precambrian rocks in the upper Kafue Basin of Zambia was investigated by Webster (1965). The site was some 40 km southwest of N'dola, where the gently undulating plateau surface had an elevation of 1200–1300 m and a local relief of 60 m over a distance of a

134

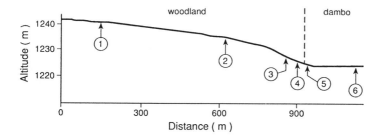

Pit No.	1	2	3	4	5	6
colour	red ——————————————→			yellow ——————→		grey
texture	sandy clays ——————————————→			loamy sands →		
silt/clay	0.4	0.1	0.4	1.2	1.4	0.9

Figure 9.4 Zambia, Webster's catena (after Webster 1965).

kilometre from a gently rounded crest to a slightly incised streamline. The vegetation was a typical central African savanna woodland, with trees giving way to grasses and sedges along the drainage lines (dambos). Rainfall averaged 1300 mm yr^{-1} with a summer maximum. A typical soil of the undulating plateau surface, midway between crest and streamline, was described as consisting of:

> ... several feet of bright concolourous friable earth with organic staining in the top few inches. This usually overlies a horizon of nodular concretionary ironstone, or gravel frequently with associated quartz stones, which in turn passes down into a mottled zone showing to varying degrees the structure of the original rock.

In the terms used in this book this is clearly a texture-contrast soil, consisting of a topsoil with a basal stonelayer over a saprolitic subsoil. Webster interpreted this in a zonalistic manner (cf. Fig. 9.12), claiming that contemporary pedogenesis was restricted to the brightly coloured fine earth above the stonelayer, and it was to this material that investigation was confined. Six pits were dug from the crest to the drainage line (Fig. 9.4) to expose a sequence of soils very similar to the type sequence described by Milne (Fig. 9.3). There was a colour change from red at the crest to yellow on the lower, well drained slopes, just above the greys of the dambo soils. In the same direction there was a change in field textures from sandy clays near the crest to loamy sands farther down slope.

135

It was postulated that the material above the stonelayer resulted from the subsoil saprolite being mined by termites. The fact that this material was redder in an upslope position and the silt:clay ratio reached a maximum on lower slopes was taken to mean that the upslope material was older (Kubiena 1956) and the downslope material was younger (Van Wambeke 1959). The uniform distribution of non-opaque heavy minerals in the fine sand fractions across all six pits was interpreted as meaning that the material supplied by the termites was then subject to a long period of *in situ* weathering and leaching to explain its present low silt content, low cation exchange capacity, low base content and the dominance of kaolin in the fine fraction.

This left Milne's concept regarding surface erosion and deposition to be resolved, for the differential removal of fines by rainwash was recognized by Webster as taking place throughout the catena, with the amount involved being related to the degree of slope. Thus, there was a minimum removal near the crest and a maximum on the lowest and steepest slopes. The solution to this dilemma was to reduce the possible pedological significance of this process by appealing to definitions and treating it as part of geomorphology. This was done by arguing that, even where the effect of differential removal was at a maximum, that is on the lower slopes, it did not extend beyond a depth of 40 cm, below which the *in situ* processes of weathering and leaching were dominant. In addition, once this material was detached it moved so quickly down slope and out of the system that there was little opportunity for it to become mixed with the soil over which it moved. In other words it acted as an independent body, which was not subject to pedological processes as it moved down slope; it was considered to be an example of mass wasting, a geomorphological rather than a pedological process. Thus, although the process of erosion and deposition was recognized, its pedological significance was denied (cf. Nikiforoff 1949).

From this discussion it can be seen that the model of soil formation used by Webster was concerned with a series of distinct successive episodes, even though only topsoil formation was being considered. First of all the topsoil was emplaced by termite activity over a considerable period of time, for the material at the crest was claimed to be significantly older than the material on the lower slopes. It was then subject to long-term weathering and leaching, while at the same time being subject to geomorphological surface mass wasting, which had no pedological significance. However, it is possible to rationalize this complex explanation in terms of the processes discussed in Part I of this book. First of all the low nutrient status of the topsoil is explicable in it being derived from the saprolitic subsoil by mesofaunal activity, where it had already been subject to strong epimorphism. Therefore, it is unnecessary to invoke a long period of *in situ* weathering and leaching after the topsoil has been emplaced. There is also no need to invoke "mass wasting" to explain surface erosion and deposition, for everything can be explained by a combination of mesofaunal activity and rainwash leading to differential lateral transport.

Another investigation was made near Harare (Watson 1964) in almost identical conditions of bedrock, topography and vegetation to those described by Webster in Zambia. The site was somewhat higher (1550 m) and the summer rainfall more restricted (840 mm yr^{-1}). Seven pits were dug along a transect from a drainage divide to a streamline (Fig. 9.5). This was a typical catenary sequence in which the threefold morphology was clearly distinguished: a sandy topsoil with a quartz stonelayer at its base, possibly mixed with ironstone gravels and boulders, or cemented by iron, underlain by saprolite. The topsoil varied in thickness down slope from a few centimetres near the crest to almost 3 m on the lower slopes. The basal surface between the fresh granite and the saprolite was sharp and highly irregular, so that saprolite thickness varied from nothing where the granite outcropped, to 11 m or more. It was apparent that epimorphism was producing saprolite *in situ*. There was also evidence in pits 1 and 2, where the topsoil/stonelayer was not too thick, that near its upper margins the saprolite was being destroyed by the burrowing activity of the mesofauna, and this conclusion was supported by the clay content of nearby ant and termite mounds, which indicated a saprolitic source. There was also abundant evidence of surface erosion and deposition. Thus, once again the same

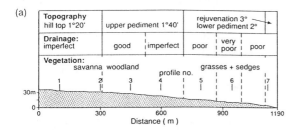

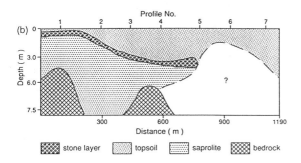

Figure 9.5 Zimbabwe, Watson's catena (after Watson 1964): (a) topography; (b) soil materials.

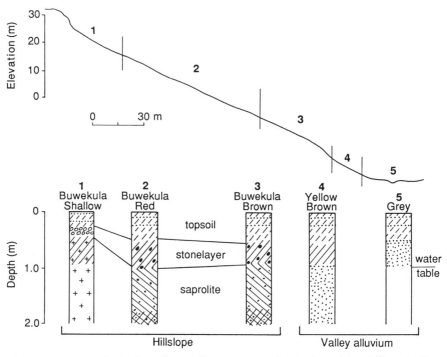

Figure 9.6 Uganda, Radwanski & Ollier's catena (after Radwanski & Ollier 1959).

processes were operating, as were discussed in Chapter 4, to produce a texture-contrast soil with a stonelayer at the base of the mobile topsoil.

The same situation was also to be seen in the investigation by Radwanski & Ollier (1959) of a catenary sequence of soils in western Uganda (Fig. 9.6). The area had an elevation of 1200 m and a local relief of 100 m over a horizontal distance of 350–400 m between hillcrest and streamline. The bedrock was again a Precambrian granite and the original forest vegetation had changed to a fire-climax savanna. This site differed from previous examples in that the slopes were much steeper (9–10°) and granite outcrops were common throughout the landscape. From the section given in Figure 9.6 of a typical hillslope, it was possible to recognize a topsoil with a basal gravel layer, which thickened to nearly 1 m down slope, over a granite saprolite. It was demonstrated (Fig. 9.7) that similar threefold morphologies were common in eastern Uganda (Ollier 1959). The formation of this threefold texture-contrast morphology was seen as being the result of a sequence of four quite separate processes:

- deep alteration of bedrock
- erosion of graded slopes across the deeply altered mass

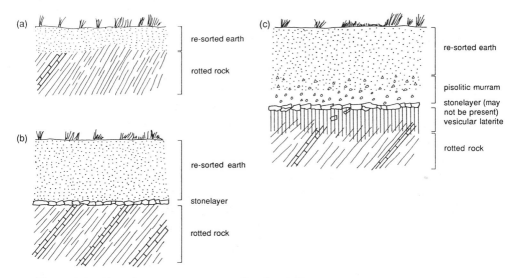

Figure 9.7 Uganda, texture-contrast soils (after Ollier 1959).

• formation of topsoil and basal stonelayer by mining of the subsoil by mesofauna and subsequent sorting also by mesofauna (such biospheric activity also explained why the topsoil and stonelayer was continuous across topographic highs)

• pedogenesis of the topsoil, so that the profile sank into the landscape at the same rate as the surface was removed by erosion (cf. Fig. 9.12).

It was an equilibrium process for non-cumulative soils (Nikiforoff 1949) in which the erosion products were removed so rapidly that they were not affected by pedological processes. Once again surface erosion was recognized and its pedological importance denied.

Despite the emphasis given to contemporary pedological processes being *in situ* and operating vertically downwards, reference was also made to "drift soils", that is soils in which the topmost layer was laterally mobile. Thus, it was stated by Radwanski & Ollier (1959):

> The only well marked drift soils occur around some fresh rock out-crops, where debris derived from the fresh rock is spread out as an apron around the rocks. Here there is a profile with a topsoil rich in weatherable minerals overlying a subsoil of rotted rock from which virtually all minerals have gone.

If downhill movement of topsoil is recognized at such sites it surely must occur at others. Indeed, when account is taken of a 9–10° hillslope, combined with surface material being sandy and having a weak fabric, it is

139

difficult to see how lateral movement is not general throughout the area, even though it may be rather more difficult to recognize where a strong mineral contrast does not occur.

In all three of these examples from East and Central Africa an attempt was made to explain texture-contrast morphology in zonal terms, but at the same time Milne's influence caused surface erosion and deposition to be taken into account. The results were highly equivocal. At the same time, we have shown that the model developed in Part I and applied to Australian soils in Chapter 8 provided a perfectly satisfactory solution.

In the West African forest zone a great deal of pedogenic work was done on similar texture-contrast soil materials. Once again the pioneer work of Milne was used as a basis, but the solution built on this foundation was highly pragmatic, rather than zonalistic, being closely associated with the problems of growing cacao. Charter (1949a,b) described a typical soil (Fig. 9.8) as consisting of: an upper light-textured, humus-stained layer, some 30 cm thick; a middle zone 60 cm to 1 m thick, of sandy kaolinitic clay with ironstone concretions and frequent quartz gravels; and a lower layer of clayey, altered bedrock, grading into fresh bedrock at a depth of 3–5 m. In other words this was a typical texture-contrast soil, consisting of a loamy topsoil with a basal stonelayer over a saprolitic subsoil.

A catenary sequence of such soils was investigated by Nye (1954, 1955a,b,c) in the Ibadan region of southwestern Nigeria. The gently undulating topography had a local relief of 40 m over about 700 m between interfluve and streamline. For the most part crests were smooth and rounded, but occasionally there was an outcrop of granite gneiss bedrock. The original high forest had been replaced by secondary regrowth. Six pits were dug distributed from the crest to near the streamline, and in all of them it was possible to distinguish a topsoil/stonelayer/saprolitic subsoil sequence (Fig. 9.9).

Nye established that the topsoil together with its basal stonelayer was sorted by the activity of the mesofauna. In addition he established that within the finer-grain topsoil it was possible to distinguish a coarser layer attributable to termites and an overlying finer layer from the breakdown of worm casts. Of even greater importance, it was shown that the topsoil and stonelayer as a whole was moving down slope whereas the subsoil remained *in situ*. It was realized that a consequence of such movement was that the zonalistic A/B/C-horizon nomenclature was inapplicable to such soils. The mechanism suggested for the downhill movement was **creep**. Despite the fact that this term was used repeatedly, nowhere was it explained what evidence was required to prove that creep had been operating; it was used rather like a "black box". Creep was postulated to operate in combination

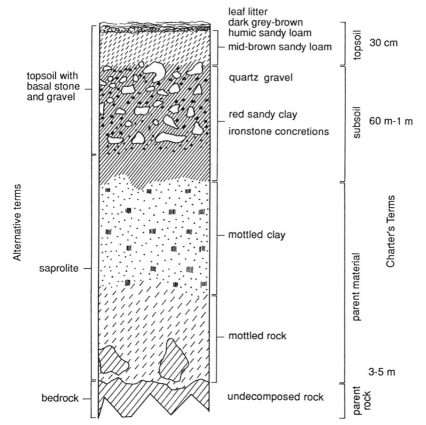

Figure 9.8 West Africa, texture-contrast soil (after Charter 1949b).

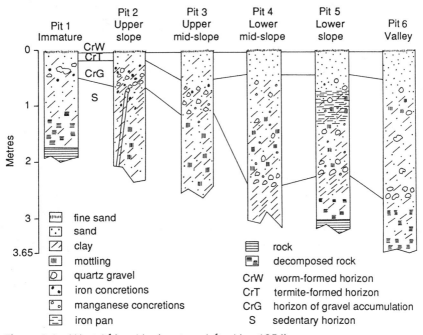

Figure 9.9 West Africa, Nye's catena (after Nye 1954).

with clay eluviation, for high soil porosity was taken to mean that all drain-age was vertically downwards, to form the sandy soils of the lower slope (pit 5). However, the experimental procedures used to demonstrate the neces-sity of vertical drainage were simplistic and hardly sufficient to establish it in an unequivocal manner. This was particularly so when it had been recog-nized by Charter (1949a,b), on the basis of much more field experience in the West African forest zone, that rainfall was at times too intense for it all to percolate downwards, no matter how porous the soil. As a result water moved laterally on and through the near-surface soil, carrying with it towards the drainage lines more of the finer particles and leaving behind the coarser particles as residuals on the lower hillslopes. In other words rain-wash was the operative process as discussed in Chapter 4. Figure 9.10 illus-trates the consequences of such lateral movement down a typical hillslope. As interpreted by Charter the topsoil in section A consisted of a mixture of all particle sizes derivable from granite, which could act as a source for materials farther down slope. In sections C and D coarse sands had accu-mulated on lower hillslopes, whereas in section E there was a concentration of finer-grain materials.

Charter also realized that the relative and absolute thickness of the top-soil, together with the stonelayer and the saprolite, could change consider-ably depending on local environmental conditions. Thus, under much greater rainfall in Sierra Leone and the Cameroons, both topsoil and sapro-lite increased considerably in thickness, but even so the basic morphology of topsoil with basal stonelayer over saprolite could still be recognized and all these soils could be explained in terms of a mobile topsoil moving down slope across an underlying saprolitic subsoil.

The work of Nye and Charter established the fact of topsoil mobility and the fundamental importance of mesofaunal activity in developing topsoil morphology and in particular that of the basal stonelayer. In effect they established a model for the development of texture-contrast soils, which was very similar to that proposed in Part I based on work in Australia. It was unfortunate that such a dynamic and integrated model was submerged by the introduction, at about this time, of a strictly zonalistic interpretation of these soils, at other localities in Africa.

This arose from work in Zaire by members of the USDA initiated by Kellogg (1949). The common soil that was encountered was a red or yellow sandy clay loam with a uniform or gradational profile to a depth of a metre or more. Fabric was generally porous earthy with little if any sign of pedal-ity. As soil profiles such materials were almost totally lacking in any out-standing characteristics, apart from some surface organic staining. In such circumstances B horizons were difficult to identify in a consistent manner.

142

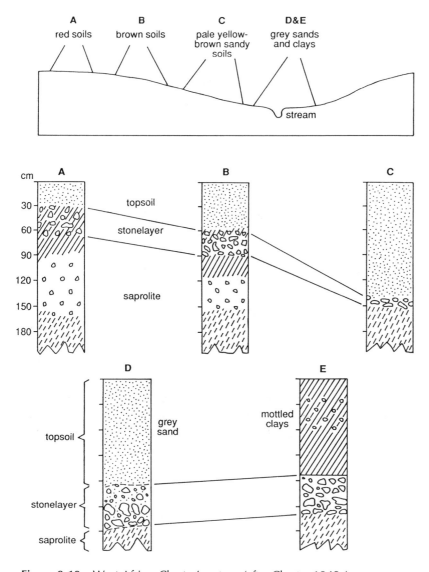

Figure 9.10   West Africa, Charter's catena (after Charter 1949a).

This was particularly so in the case of its lower boundary, and in most cases it was no more than vaguely located in the region where plant roots died out. Beyond this point the same kind of material continued on without any real change, and yet according to zonalistic ideas this had to be the material from which the solum (A and B horizons) developed. It was this type of situation that caused the USDA to develop the concept of parent material, as discussed in the Introduction. Such parent material was seen as having been devel-

143

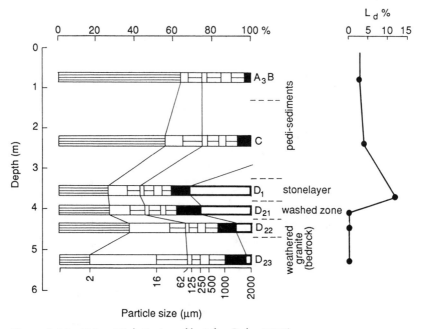

Figure 9.11   Zaire, High Ituri profile (after Ruhe 1956).

oped from parent rock elsewhere before being moved to its present situation as a sediment. It was then subject, over a long period of time, to the effects of the active factors of soil formation, which resulted in the present deep uniform soil of low fertility. Even though uniformity was generally taken as a sign of immaturity, Kellogg (1949) regarded these soils as being the end-point of development, and hence they were mature soils. In addition, in view of their wide distribution and long exposure to the active factors of soil formation, they were interpreted as being the zonal soils of the tropics and were given the name **latosols**. This initial interpretation was extended and refined by Kellogg & Davol (1949), Ruhe & Cady (1955) and Ruhe (1956). At the same time and particularly in the last case, where the work was done in the northeast of Zaire, it was seen that the deep uniform latosol was underlain by a saprolitic bedrock, and at the junction between the two there was generally a well developed stonelayer (Fig. 9.11). As the top half of the Zaire sections had already been interpreted as a profile in restricted zonal terms, the rest of the section had to be interpreted in a similar manner. The necessary genetic sequence in zonalistic terms (Fig. 9.12) consisted of:

1. deep alteration of bedrock
2. erosion to leave a saprolite covered by a lag gravel
3. sediment, generally regarded as being Tertiary in age, deposited on top of the lag gravel, converting it into a stonelayer

144

4. sediment acting as a parent material subject to long-term pedogenesis to form a latosol. This made the residual saprolite a **paleosol**.

The fundamental weakness of this explanation was in its complexity, for it was made up of a series of contingent events. The chance of these events occurring in this particular sequence even once was remote and yet, as this morphology was very common, what was being demanded was a repetition of the sequence many times over or a situation where such a sequence of chance events extended uniformly over a great area. This particular approach strongly influenced the work of Webster (1965), Watson (1964) and Radwanski & Ollier (1959) discussed earlier in the chapter.

Of even greater significance, however, was that these Zairean soils had the same morphology as those previously discussed in East, Central and West Africa. What were being dealt with throughout were texture-contrast soils with a thick accumulative topsoil. All of their features could be explained in terms of the unified integrated scheme involving epimorphism and near-surface processes as discussed in Part I and Chapter 8, which avoids using a complex sequence of contingent events.

However, the zonalistic style of interpretation had very strong taxonomic links, which can be seen in the eventual absorption of the latosol concept into *Soil taxonomy* (USDA 1975). There was a strong demand for such a scheme, in terms of mapping and classification to serve more pragmatic agricultural ends, and, because of this, its use spread rapidly, particularly among soil taxonomists, throughout Africa (Sys 1960, Duchaufour 1982) where, despite its inherent weaknesses, it has to date swamped alternative possible genetic explanations.

It can now be seen that as a result of Africa being a lithospheric plate centre, where the determinative factors of soil formation are a granitic bedrock

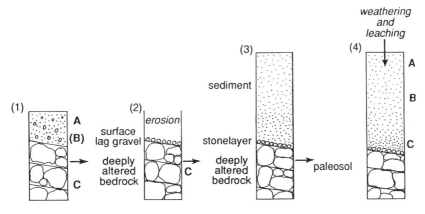

Figure 9.12 Deep texture-contrast soil zonal interpretation.

and minimal topographic variation, and where volcanism and seismicity are of little importance, the modal processes of epimorphism and near-surface transport tend to produce a texture-contrast soil generally with a deep accumulative topsoil, as a normal endpoint. The fact that Africa has been a landmass for hundreds of millions of years means that these pedogenic processes have been operating continuously over a very extended period of time, which explains the accumulation of the masses of quartz sand, the residual endpoint of pedogenesis, not only in the Kalahari, the Congo Basin and the Sahara but also throughout a great deal of Central and West Africa. In other words the situation is very much a parallel to that which occurs in Australia.

In the next chapter it is proposed to examine how much further this extrapolation can be extended, to other plate centres.

CHAPTER 10

Soils of other
continental plate centres

Any adequate account of scientific method must include a theory of
incentive or special motive: must contain a canon to restrict observation to something less than the whole universe of observables. We cannot browse over the field of nature like cows at pasture.
Peter Medawar

Across the South Atlantic on the high plateau of southern Brasil texture-
contrast soils occur that are similar to those described in Chapter 9, which
is to be expected for this is another Gondwana fragment of a continental plate
centre with much the same range of lithospheric materials and topography.
Thus, in the region of Brasilia on a plateau surface (1000–1200 m elevation)
under a savanna vegetation, Macedo & Bryant (1987) described a sequence
of soil materials consisting of 6–9 m of uniform topsoil, with a basal stonelayer, over 5–6 m of saprolitic bedrock derived from a complexly folded
sequence of Precambrian shales, phyllites, quartzites and gneisses. This
sequence was explained as being the result of a Cretaceous to mid-Tertiary
planation, with sediment being deposited on the planation surface in the late
Tertiary. Following regional uplift at this time deep weathering of the sediment was said to have occurred before final dissection in the Quaternary to
give the present residual plateau landscape (Fig. 10.1). The present soils,
which were considered to have developed in the Tertiary sediment, were
classified as oxisols (USDA 1975) or latosols and laterites in Brasilian terms.
This interpretation was similar to that proposed by Ruhe (1956) for soils of
the Congo Basin (Fig. 9.12). As before, however, all of the criticisms levelled
at Ruhe's zonalistic explanation were applicable to this Brasilian example.
Once again the greatest equivocation was shown with regard to the nature
of the so-called Tertiary sediment. This can be seen in the description of a
similar material from a plateau surface of somewhat lower elevation
(750–800 m) farther to the south in the São Paulo region (Moniz et al. 1982:
1236; Moniz & Buol 1982) where the following comment is made:

147

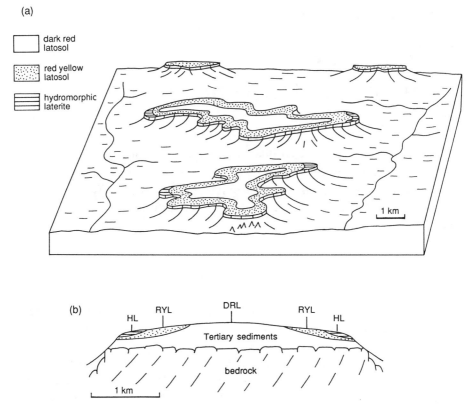

Figure 10.1 Brasilia, landscape and soils (after Macedo & Bryant 1987): (a) distribution; (b) cross section.

The soils studied are derived from a transported material similar to the underlying altered *in situ* saprolite. The material appears locally transported; however, it is probably a mixture of several materials, in varying proportions from different sources and is not recognizable and traceable because of its highly weathered condition. How many pedological cycles the material underwent prior to its last deposition and how this could affect the present soil derived from it is a matter of speculation.

In line with the strictly zonalistic interpretation of these soils contemporary pedogenesis was restricted to the movement of materials in solution, whereas the lateral movement of solids was excluded from consideration as being non-pedogenic. Yet, such processes were shown to be an integral part of soil formation on the northward extension of the Brasilian plateau in northeastern Mato Grosso (Askew et al. 1970, Townshend 1970). In this forested region the topography consisted of gently undulating plateaux surfaces, and the bedrock was composed of almost flat-lying quartzose sandstones, shales and mudstones of Devonian–Carboniferous age. The sandstones formed the top of the sequence and hence were the main source of soil material on the plateau surface, with the shales and mudstones being confined to the valleys (Fig. 10.2). On the plateau surface a topsoil 2–4 m

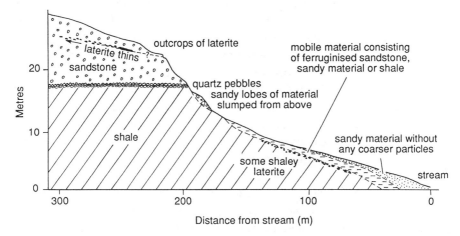

Figure 10.2 Mato Grosso, plateau/stream section (after Townshend 1970).

thick overlies a saprolitic sandstone subsoil, with the junction between the two being marked by a zone of ironstones (i.e. a stonelayer). At any given site the topsoil was more or less uniform, except for a surface zone of some 40 cm, which was less brightly coloured and more coarsely textured. Laterally, however, the topsoil displayed considerable variation (Fig. 10.3). Over the gently rounded crests the texture was a sandy clay loam, on midslopes a sandy loam and on lower slopes a loamy sand to sand, just above the clay-dominated hydromorphic soils of the drainage depressions. The colour also varied from reddish at the crest to yellowish midslope and pale brown on

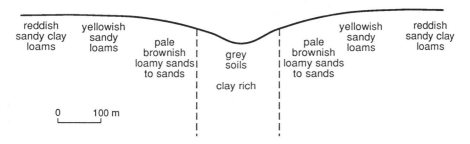

Figure 10.3 Soil variation, Mato Grosso plateau (after Askew et al. 1970).

149

lower slopes. When the amplitude from drainage line to interfluve was restricted, the yellow colour extended to the crest. This situation was almost directly comparable to that described in Zambia (Webster 1965) and in Tanzania (Milne 1947) and can be similarly interpreted as being the combined result of bioturbation and rainwash causing erosion, downhill trans- port, sorting and deposition, as described in Chapter 4. The reduction in the range of colours similarly pointed to the accumulation of coarse sand on the slopes leading to a gradual burying of the topography. In addition in this case the forest cover had given enough stability to the surface, so that the consequences of the preferential removal of the fines as a result of bioturba- tion and winnowing were observable in the surface 40 cm. If topsoil mobility can occur under the forested conditions of the Mato Grosso, it would be even more likely to occur farther south under the savanna vegetation in the region of Brasilia and São Paulo. However, in these areas the pedological approach has been strictly zonal and processes of lateral surface movement have been excluded from consideration and, hence, have not been dis- cussed.

It would seem, from the evidence available, that the processes of soil formation on this stable South American block are the same as those that have been described as operating in Africa and Australia, which give rise to texture-contrast soils, many of which have deep accumulative topsoils. The whole of the discussion in this and the previous two chapters suggests that these soil processes and resulting soil materials are a common feature of Gondwana fragments, and therefore both processes and materials can be extrapolated to other Gondwana fragments where similar conditions prevail such as Madagascar, Ceylon and peninsular India.

If such an extrapolation is possible, the next question must be to what extent can this be extended into the northern supercontinent of Laurasia, which has had much the same history of fragmentation as Gondwana and consists of equally broad areas with bedrock and topography typical of con- tinental plate centres. An essential step in this direction was provided by Berry & Ruxton (1959), who realized that sections in Hong Kong consisting of 50–60 m of saprolitic granite overlain by 4–5 m of topsoil with a basal stonelayer were comparable both in morphology and in mode of formation with the much thinner sequence described by Ruxton (1958) on a Sudan granite, which was discussed at the beginning of Chapter 9. In other words they were both explicable in terms of a mobile topsoil with a basal stonelayer moving down slope across a saprolitic subsoil. Comparable soils to those of Hong Kong are developed on the granites of the Malaya peninsula and southern Thailand, but are largely unrecognized because of lack of expo- sures. However, there are occasional deep sections resulting from quarrying and road-making where the characteristic morphology can be recognized within a total thickness in excess of 20 m to fresh bedrock (Plates 7 and 8). A similar section was described by Eswaran & Bin (1978) overlying granite bedrock near Kuantan on the east coast of Malaya. The total depth was 19 m, of which 16 m was saprolite and 3 m topsoil. The morphology was recognized as being so similar to that of soils in Zaire (which were discussed in the last chapter) that the nomenclature developed there by Stoops (1967) was applied to the Kuantan soil: α = topsoil, β = stonelayer, γ and δ = saprolite. It is probable that texture-contrast soils of this type are widespread in this region of Asia.

In considering the continental plate centres of northern Laurasia in both Eurasia and North America, account has to be taken of a fundamental change of conditions compared to those that occur at the other plate centres, for there are large areas where temperatures are below 0°C and the ground is frozen. In other words such areas are subject to **permafrost**, where cryoturbation dominates. Within such regions a unique set of near-surface processes are operating, which are clearly non-pedogenic. Such effects are very widespread (Fig. 10.4) for they occur not only in the area permanently subject to permafrost, but also in those areas where permafrost is discontinuous.

Account also needs to be taken of the fact that northern Laurasia was repeatedly glaciated during the Pleistocene (Fig. 10.5), for as a result great areas have been covered by the products of glacial and periglacial processes. The most important thing about these materials is that none of them have been exposed to pedogenic processes for longer than about 10 000 to 12 000 years, since the retreat of the last ice-sheet. From what has been said previously about the time required for soil development it can be concluded that within such a time period not a great deal of pedogenesis is to be expected, and that which does occur will be the result of near-surface processes rather than epimorphism. It is unfortunate, therefore, that the great majority of soils investigated in western Europe and eastern North America have been cultivated, which has destroyed a great deal of the evidence of the operation of near-surface processes. This situation is exacerbated by the fixation on profile description and the rejection of anything connected with possible lateral movement. Thus, despite the lead provided by Darwin's (1881) study of earthworms, when the importance of a mobile topsoil was clearly recognized, little attention has been paid to such features in the northern hemisphere apart from comments by Robinson (1936) and more recent work in the Ardennes by Imeson and his group (Imeson 1976, 1977, Imeson & Kwaad 1976, Imeson & van Zon 1980, Kwaad 1977) and Johnson (1989, 1990) in the USA.

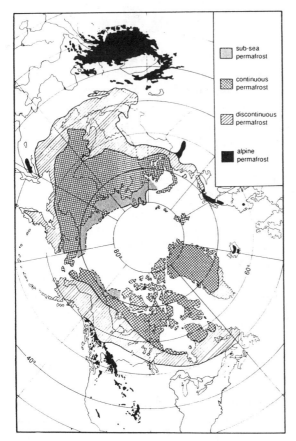

Figure 10.4 Permafrost, northern hemisphere (after Péwé 1991).

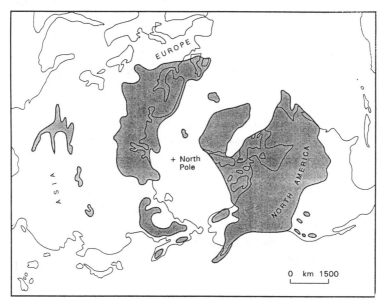

Figure 10.5 Extent of Pleistocene ice-sheet (after Price 1973).

It can be expected that the character of the soil material resulting from glacial and periglacial processes must be dominated by inheritance and, therefore, it is necessary to consider the nature of these materials in some detail. This is best done in two parts; first of all for those areas that were formerly covered by ice-sheets, and then for those areas associated with the end moraine complex and outwash zones beyond the ice limits. In moving across the landscape, ice-sheets remove surficial materials, as well as grinding and eroding fresh bedrock in preglacial hilly and mountainous areas and dumping the debris in former valleys. The melting of the ice-sheet exposes bare rock surface interspersed with boulder clay plains, neither of which is conducive to pedogenesis. On bare rock surfaces water tends to run off rather than penetrate and hence epimorphism is restricted. A similar situation exists on the boulder clay plains, for again water does not penetrate easily so that surface-water gleys predominate, in which inheritance is much more important than any form of epimorphism. The boulder clay is of course deposited along the preglacial drainage lines, so that when the ice melts the natural drainage lines are blocked and considerable areas of swamp develop.

The limit of the ice-sheet is marked by an end moraine complex where all the debris carried by the ice-sheet is finally deposited. Across the area of moraine deposition, which is bare of vegetation, there is an abundance of meltwater and strong winds generated by the high atmospheric pressure over the ice, so that conditions are highly conducive to lateral surface movement. However, the materials that are affected in this way are physically and chemically different from the materials considered in Chapters 4 and 5, which had been subject to epimorphism. In the glacial terrain, material has been comminuted by grinding, which produces a continuous spectrum of particle sizes, compared with the bimodal particle-size distribution resulting from epimorphism. In particular this means that silt-size particles are produced and retained, whereas under conditions of prolonged epimorphism they are eliminated. In addition during rock grinding the lattices of the cation-rich minerals (the mafics and feldspars) are easily distorted or shattered and show a greater tendency than quartz to become silt-size. Furthermore, the lattice distortion means that the associated cations become more easily available for plant growth, which has considerable additional pedological implications.

The morainic debris is first of all affected by seasonal glacial meltwaters, which give rise to extensive outwash plains and braided river channels, made up for the most part of fine gravels, sands and silts, for the larger clasts are left behind on the moraine and the clay is transported farther down stream before it is deposited. The silt is removed from these deposits by the strong turbulent winds associated with the boundary of the ice-sheet. A combination of these fluvial and aeolian processes leads to a segregation of sand-size particles where quartz is overwhelmingly dominant and now covers extensive areas in northern Eurasia and North America. Pedologically these quartz sands behave in exactly the same way as quartz sand accumulations in humid parts of Australia (see Ch. 8) and for the same reasons podzols result.

The wind-transported silts are not dispersed over such a wide area as finer clay-size particles (see Ch. 5) and this is reflected in the discrete accumulations of **loess** that have resulted from the Pleistocene glaciations, which occur in a broad latitudinal band across Eurasia and North America (Fig. 10.6; Pye 1987). These deposits give rise to soils of high fertility,

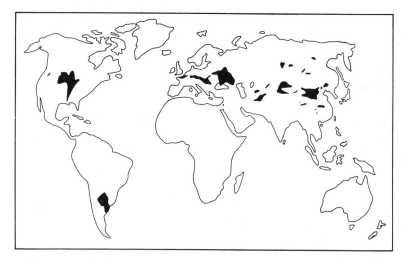

Figure 10.6 Major areas of loess deposition (after Pye 1987).

because of their glacial inheritance, which has not been affected subsequently to any extent by pedogenic processes. This is seen in the case of the Ukrainian **chernozem**, which has developed on the extensive loess deposits of the region, under native grasses, merely by the accumulation of a thick organic-rich topsoil on and within the surface of the loess. This same dominance of inheritance is also seen where the loess has been redistributed by fluvial processes. Thus, Ruhe (1984a,b) examined 25 loess-derived soils spread over an area from Minnesota to Mississippi and showed that all soil properties were explicable in terms of the depositional system that laid them down. In other words in the 8000 or so years since they were deposited there is little evidence of pedogenesis.

Thus, these continental plate centres of northern Laurasia, which have been subject to Pleistocene glaciation, present a totally different pedological picture from those of the other plate centres previously discussed, for the recent deposition of a new lithospheric material means that inheritance is dominant, as compared to the advanced state of pedogenesis that occurs at other plate centres. The short time over which pedogenesis has been operating on this glacially derived lithospheric material is also reflected in podzols and chernozems being the outstanding soils of the region, for both can develop in a very short time. Thus, podzols only require the stripping of the thin oxide-rich covering of the quartz grain by chelation and its deposition lower down to form pans, which is easily accomplished in a quartz sand (Paton et al. 1976). In the case of the chernozem, all that is required is the accumulation of a surface layer of well humified organic matter, together with the movement and concentration of calcium carbonate at a greater depth. In view of the abundant growth of grass and the highly porous nature of the loess, this need not take very long, and it has been suggested that a few centuries would be sufficient (Flint 1971). In these cases there has been little fundamental change in either the quartz sand or the highly porous loess, which is very different from the fundamental processes of reorganization required to produce the texture- and fabric-contrast soils, so characteristic of materials exposed to long-term pedogenesis at the other continental plate centres.

It can, therefore, be concluded that northern Laurasia has a markedly aberrant style of soil formation compared to continental plate centres elsewhere in the world. Yet, it is from within this region – the plainlands of European Russia and the Midwestern plains of North America – that the zonalistic model of soil formation was developed and then applied to the rest of the world. The consequences of this are well seen in the case of the podzol, which, with its distinctive brightly coloured profile standing in marked contrast to the general drabness of the surrounding gley soils, caused it to be recognized as a zonal soil characteristic of cold, wet regions with an evergreen vegetation. In addition, because of the accident of its early discovery and prominence, the pedological importance of podzols was extended far outside these northern regions and in fact became worldwide, for the way podzols developed by vertical translocation was taken to be the model for soil or profile development throughout the world and led to the adoption of A/B/C-horizon nomenclature with its implicit vertical genetic connection for all profiles. It also emphasized that B horizons resulted from the movement of material from the surface and its deposition. The seriousness of such developments can be gauged when it is realized that podzol development depends absolutely upon the segregation of largely inert material (i.e. quartz sands) under humid conditions, which is a very special and restricted set of circumstances, compared to the generally applicable model of texture-contrast soil formation, developed in this book, where A/B/C-horizon nomenclature is inapplicable and the *in situ* subsoil (the so-called B horizon) is to a large extent formed by inheritance from lithospheric material and not by translocation of material from the surface.

It can also be recognized that the original zonal distribution of soils within such glacially affected regions, from podzols in the north to chernozems in the south, which was explained as being the result of climatic influences, is, now that inheritance has been more clearly defined, more cogently explicable as resulting from the fluvial and aeolian redistribution from an end moraine complex, where the contribution of pedogenesis remains slight. If this conclusion is accepted, it follows that these northern areas of Laurasia, rather than being accepted as type areas for pedogenesis, should be regarded as a special case where there has been little time for pedogenic development. At the same time it must be accepted that the type areas for pedogenesis are located on continental plate centres that are not at present subject to permafrost, or have been subject to glaciation in the Pleistocene. In other words the type areas for pedogenesis are to be found in Australia, Africa, cratonic South America and peninsular India, and it is from such an area, in southeastern Australia, that the discussion of pedogenic processes, which forms Part I of this book, was derived.

It is rather ironic that such a claim as that in the last paragraph receives support from evidence that occurs within North America. In the case of this continent the southern limit of glacial advance has been precisely demarcated. Immediately to the south of this boundary are areas that were protected from the effects of the glaciation, such as the Piedmont region in the lee of the Appalachians and the Ozark Plateau farther to the west, where pedogenesis was able to continue throughout the Pleistocene. Within such regions it can be recognized that what is being dealt with is a continental plate centre with normal lithospheric material and topography, which has been exposed to pedogenic processes for an extended period of time. In other words this setting is equivalent to that which has been discussed for the much more extensive continental plate centres in this and the previous two chapters and, as would be expected, the resulting soil material in these two areas is the familiar topsoil with basal stonelayer over a deep saprolitic subsoil (Hunt 1966). The zonalistic influence intrudes even here, for these soils were recognized as being red/yellow podzolics, the zonal soils of the humid subtropics. The term "podzolic" was taken to signify that podzolization was occurring, which was reflected in the translocation of clay to form the B horizon (see Introduction) but had not yet progressed to the translocation of sesquioxides and the formation of true podzols. This same pod-zolic nomenclature had also penetrated and been applied to similar soils in southeastern Australia (Stephens 1962), which are of course the texture-contrast soils described in Chapter 8 (Table 8.1).

A correlation between these North American "islands" and other conti-nental plate centres was first suggested by Milne (1939) during a visit to the USA, when he recognized a general similarity between the soils of the Pied-mont and those of East Africa with regard to the topsoil and the saprolitic subsoil, and more particularly with the stonelayers at the topsoil/subsoil junction. His visit coincided with the publication of *Landslides and related phenomena* by Sharpe (1938) in which for the first time stonelayer data, almost all from the Piedmont, were interpreted as indicating that stone-layers formed along the base of the topsoil, which was mobile and moving down slope (Ireland et al. 1939). The idea was enthusiastically adopted by Milne as explaining a great deal about East African soils. However, it was later rejected in the USA, particularly with regard to the formation of stone-layers (Parizek & Woodruff 1956, 1957a,b), which were regarded as having been formed initially as a surface lag gravel. This interpretation was endorsed on strictly zonalist grounds (Fig. 9.12) by Ruhe (1959) and such interpretations continue (Darmody & Foss 1982). Yet, from the arguments put forward in this book, it would seem that Milne's correlation was well founded, and indeed the occurrence of such "islands" of texture-contrast soils, sheltered from the effects of the Pleistocene glaciations, suggests that the model of soil development, presented in Part I of this book and applied to continental plate centres here in Part II has worldwide applicability.

CHAPTER 11

Soil materials of continental plate margins

I have an old belief that a good observer really means a good theorist. *Charles Darwin*

As previously stated in Chapter 7, lithospheric plates have both tensional and compressional margins (Table 7.1). Tensional margins are pedologically less significant for they generally coincide with the mid-oceanic ridge, which is usually submerged beneath the deep ocean. It is only in exceptional circumstances, such as when a tensional margin crosses a continent, that basic rocks appear on the Earth's land surface to be affected by pedogenesis. The East African Rift is a modern example but it represents incipient rifting only and associated basaltic eruptions, which are not very extensive, are confined to the Ethiopian Highlands. Extensive terrestrial basaltic lava flows were associated with the gradual fragmentation of Pangaea (Fig. 11.1), for this entailed major tensional rifting across the supercontinent and, where this was accompanied by overheating of the mantle, flood basalts resulted, which covered large areas on either side of the rift (White & McKenzie 1989). Three such episodes of flood basalt eruption can be distinguished, which at a global level are of pedological significance. The first of these occurred when Antarctica broke away from the southeast coast of Africa 170 Myr ago and formed the basalts of Natal and surrounding areas. The second episode was associated with the opening of the South Atlantic between Africa and South America 120 Myr ago, where the basalts now cover a large area in southern Brasil and extend into Uruguay. The third episode occurred some 65 Myr ago and resulted in the eruption of the Deccan traps, which cover much of northwestern peninsular India.

Of much less significance in terms of area are the volcanic oceanic islands that occur where localized hotspots have broken through the lithospheric plate and the resulting sea-floor volcano has been built up at such a rate that it reaches the ocean surface. When such a phenomenon is combined with lateral plate movement, an island chain develops, of which the most famous

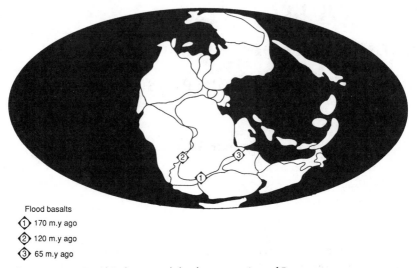

Flood basalts
① 170 m.y ago
② 120 m.y ago
③ 65 m.y ago

Figure 11.1 Basaltic lavas and the fragmentation of Pangaea.

example is the Hawaiian islands, although there are several other examples in the Pacific basin. Even though these oceanic islands are far removed from tensional lithospheric plate margins, it is convenient to include them under that heading for all the environmental conditions given in Table 7.1 apply to these small areas too. However, the same cannot be said for the three major flood basalt areas of Natal, southern Brasil and the Deccan, for, since the time of their eruption, they have moved away from the locus of volcanicity and seismicity associated with the rift and now form part of the trailing edges of continental plate centres, and hence these areas have become subject to the same environmental conditions as plate centres (Table 7.1), except that basalt is the bedrock.

The soils resulting from basaltic pedogenesis have already been discussed in Chapter 8 (Fig. 8.7). Epimorphism of basalt under conditions of good drainage produces a red, brown or yellow clay, consisting of kaolin physils with oxides and hydroxides of iron and aluminium, which generally possesses a certain degree of subplasticity. Late-stage hydrothermal alteration of the basalt results in the formation of dark-coloured, plastic, smectite clays, together with abundant calcium carbonate, which directly forms soil material by inheritance rather than as a result of epimorphism. Both the strongly coloured subplastic and the dark-coloured plastic clays are affected by near-surface processes, particularly those of the biosphere, to produce a mobile topsoil with a distinct fabric. In other words a fabric-contrast soil is produced.

Compared to the largely submerged tensional margins of lithospheric plates, compressional margins are above sea level. Indeed, they form the world's great mountain belts, peripheral to the dispersed continental lithospheric plate centres of Laurasia and Gondwana, around the Pacific and through Asia to the Mediterranean (Fig. 7.4). Given these circumstances it is apparent that the broad environmental controls are very different from those that pertain at plate centres (Table 7.1) and this is best seen in terms of the determinative factors of soil formation: lithospheric materials and topography. Commonly lithospheric materials at compressional lithospheric plate margins are fine-grained and not too well sorted sedimentary rocks, which form such steep slopes as to cross the threshold of stability, when affected, as they commonly are, by seismic disturbances. This results in frequent mass movement in the form of rock falls, landslides and debris flows, which can vary from being highly localized to those that cover many square kilometres. The most important pedological consequence of such instability is that lithospheric material is moved at such a rate that little time is available for much pedogenesis to occur, so that the character of the resulting material is almost totally determined by bedrock inheritance.

This situation occurs in the central highlands of New Guinea where, in extremely steep topography, the bedrock is a fine-grain mudstone, which has been shattered into gravel-size fragments as a result of active tectonism. In the surface zone such materials form a **diamicton** layer overlying the *in situ* mudstone. On the prevailing steep slopes this makes for a very unstable situation no matter what the vegetational cover, so that mass movement is very common, with failure occurring anywhere within the diamicton layer and even within the *in situ* mudstone. Repeated episodes of mass movement can result in thick deposits, which may extend up slope as far as the ridge crests. However, even on such steep slopes there are places where mass movement is not completely dominant, for, given even a minimum of surface stability and where both earthworms and the products of organic matter breakdown are abundant, earthworms are able to combine the finer-grain matrix material, associated with the diamicton, with the humified organic matter to form highly stable water-absorbent casts, which in turn become peds. This results in the formation of a biomantle up to 50 cm thick overlying a diamicton layer, which when subjected to landslides tends to move as intact units. Apart from some minor sloughing at the edges, the materials within such slip blocks remain undisturbed and the surface vegetation continues to grow. The resulting scars are gradually filled by a combination of debris flows and sloughing from the head-wall scarp, which finally results in a slope carrying a mosaic of biomantle "islands" surrounded by debris flow "streams".

A similar situation is to be seen in the steeplands of the South Island of

161

New Zealand where the highly fractured schistose bedrock breaks down to form a diamicton, which is subject to downhill mass movement as debris flows. In this situation there is an abundance of organic matter breakdown products resulting from the decomposition of alpine grasses under the prevailing cold, wet conditions, which the abundant earthworms combine with the fine-grain diamicton matrix to form a highly stable, water-absorbent biomantle some 40–50 cm thick. Figure 11.2 shows a 50 cm thick biomantle developed on a 28° slope, on phyllite, at an elevation of 930 m on the Southern Alps of New Zealand. The biomantle is almost stone-free, crumb-structured, yellowish brown, sandy clay loam with abundant worm casts. It is clearly demarcated from a yellowish brown-grey diamicton 30–40 cm thick, which overlies the shattered phyllite bedrock at about a metre.

A less extreme situation occurs on the moderate to steep hillslopes of the more tectonically stable, seasonally wet and dry, Port Moresby region of New Guinea. Figure 11.3 is a typical section, where the steeper slope is

Figure 11.2
Biomantle, New Zealand steeplands.

28°

0   V = H   40 cm

(sketched from photograph by P. B. Mitchell)

① Biomantle / topsoil   ② Subsoil   ③ Phyllite bedrock

162

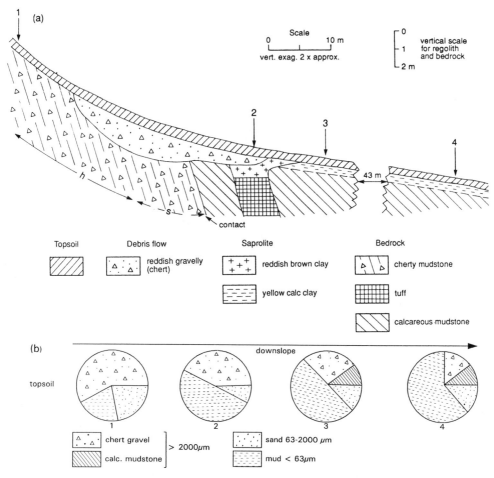

Figure 11.3 Mass movement and rainwash in the Port Moresby area.

composed of a non-calcareous cherty mudstone, whereas the more gentle slope is formed on a pale calcareous mudstone, except for a small outcrop of tuff. The steeper slope is very similar to that previously described for the central highlands, for here immediately over the bedrock is a large hollow, which was possibly created by mass movement *en bloc* and then filled by diamicton debris flows derived from the cherty mudstone bedrock. The greater stability of the lower-angle slope and the consequent greater time available for pedogenic processes is reflected in the development of saprolite in relation to both the calcareous mudstone and tuff. The topsoil is up to 50 cm thick and consists of cherty gravels and a range of fine-grain materials. The pie diagrams (Fig. 11.3b) show a downslope decrease in gravels and an increase in fines (<63 μm), which points to a derivation from the cherty mudstone and downslope redistribution by rainwash. However, this

163

does not account for the persistence of the cherty gravels at sites (3) and (4), for they cannot have been moved down slope by rainwash and must be ascribed to shallow debris avalanches from steeper portions of the hillslopes, which are a common feature of the area. Similarities can be drawn between this site and certain localities in both the humid forested mountains of northwestern USA and the drier mountains of central California, where infilled landslide scars and hollows on steep slopes are also well known (Dietrich et al. 1982, Reneau & Dietrich 1990).

The example from Port Moresby clearly shows that with a decrease in slope angle there is a decrease in mass movement and an increase in the influence of the processes of epimorphism and rainwash. In other words there is a gradual movement towards those processes that are more typical of lithospheric plate centres. This is seen particularly well in that part of New Zealand where more gentle slopes occur. However, there is one major difference, for fabric-contrast soils rather than texture-contrast soils are dominant. The main reason for this is that there are few lithospheric materials that contain sufficient residual quartz to allow the effects of differential sorting by rainwash to occur. In addition, under most forests the density of litter layers and ground cover is so great that only a very small proportion of surface earthworm casts (the dominant bioturbator) are exposed above the litter to rainwash, and furthermore surface vegetation cover is only rarely removed by wildfire. However, there are occasions when quartz-rich bedrock occurs on low-angle slopes and this results in the development of texture-contrast soils just as in Australia. Figure 11.4 shows such a situation down a 10° hillslope near Auckland that cuts through Miocene marine sediments consisting of conglomerates, claystones and fine-grain sandstones. In this situation a texture-contrast soil is developed with a silty loam topsoil over a clay saprolite with a columnar or prismatic structure. Plate-like ironstones derived from thin ironstone outcrops give rise to a sporadic stonelayer at the base of the topsoil for some 5 m down slope from their point of outcrop. In other words all the processes are very similar to those described for continental lithospheric plate centres.

In a rather similar way texture-contrast soils are to be found on the Fly Platform of New Guinea (Fig. 11.5), an extensive alluvial plain covering more than 200 000 km² to the south of the main cordillera (Schroo 1964, Bleeker 1971). This Pleistocene alluvium consists of horizontal well stratified units of gravel, sand, silt and laminated clays (Blake 1971, Blake & Ollier 1971). The undissected southern part of the platform consists of broad crests with adjoining depressions throughout which texture-contrast soils prevail with a gravel-free loamy sand to loam topsoil, overlying a more clayey stratified subsoil with a variable amount of gravel (Bleeker 1983:

ironstones
conglomerates
claystones
fine-grained sandstones
silty loam topsoil
stonelayers
columnar clay saprolite

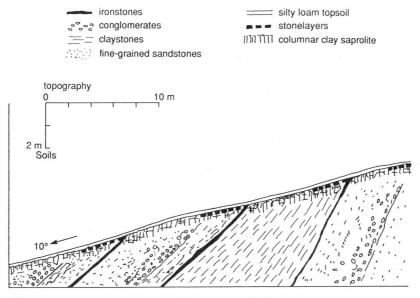

Figure 11.4 Texture-contrast soil in New Zealand.

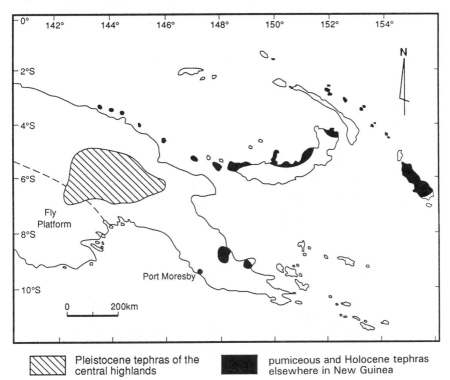

Pleistocene tephras of the central highlands

pumiceous and Holocene tephras elsewhere in New Guinea

Figure 11.5 Tephra sheets, Papua New Guinea (after Wood 1987).

145). In this savanna environment rainwash and termite activity have been noted and it is possible that these types of processes have led to the development of a biomantle at this site. It needs to be stressed that the examples from Auckland and the Fly Platform are both located away from the active plate margin and this stability combined with more gentle gradients increases the time during which pedogenic processes discussed in Part I can operate.

Table 7.1 also shows that another characteristic of compressional plate margins is explosive volcanism and the resulting deposits add yet another level of complexity to these margins in terms of the type of lithospheric material and its distribution. It is well established that the spread of volcanic debris from an eruptive centre is broadly dependent upon particle size, density, launch velocity, prevailing winds and release height (Wilson 1972). In pedogenic terms by far the most important size fraction is volcanic ash or tephra. Individual tephra deposits thin with distance from the source and their marked downwind tail gives them an overall elliptical shape (Blong 1982). Deposits at least 10 cm thick are common within 100 km of a volcano. However, repeated eruptions can lead to the accumulation of tephras many metres thick that extend for hundreds of kilometres around the eruption centre and blanket the landscape. Figure 11.6 is a typical example of a tephra sheet sequence from the North Island of New Zealand (Gibbs 1980). Similar centres occur around the Pacific Rim and includes the northwest of

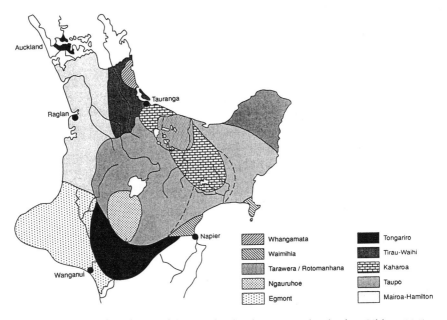

Figure 11.6 Tephra sheets of the North Island, New Zealand (after Gibbs 1980).

166

the USA, Japan, Papua New Guinea and the Indonesian islands. Tephra mantles are largely inert and a good deal of this character can be attributed to late-stage hydrothermal activity, which is commonly associated with eruptions of tephra and is responsible for the formation of considerable quantities of secondary minerals. This mineral inertness is reflected in a great deal of fabric inheritance in the tephras, such as that which occurs in the vicinity of Mt Ruapehu in New Zealand. Figure 11.7 shows six distinct

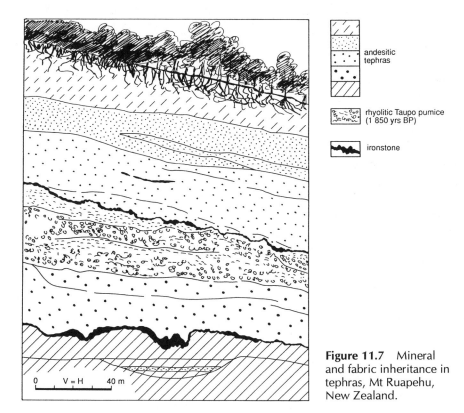

andesitic
tephras

rhyolitic Taupo pumice
(1 850 yrs BP)

ironstone

0   V = H   40 m

Figure 11.7 Mineral and fabric inheritance in tephras, Mt Ruapehu, New Zealand.

units spanning about 2 m that have been deposited in the past few thousand years. In this alpine heath environment, pedogenesis has been restricted to a slight movement of iron, which has been redeposited along the junctions between the tephra units. Tephra beds of recent age found along the north coasts of New Guinea and New Britain (Fig. 11.5) show similar fabric inheritance despite being located in humid tropical lowlands. Even in the case of much older tephras such as those of the central highlands of New Guinea, which are more than 50 000 years old (Pain & Blong 1979; Fig. 11.5), subsoils still display strong fabric inheritance, even though topsoils have been reorganized by the incorporation of organic matter by earth-

167

worms to form a thick dark biomantle. In effect a fabric-contrast soil has been formed.

It is apparent from this discussion that the soil materials of compressive lithospheric plate margins constitute a unique assemblage in which inheritance plays a dominant role. In places this uniqueness was realized at an early date. This was particularly so in the case of New Zealand when Taylor (1949) recognized that soil resulted from the interaction of three regimes: those of wasting (epimorphism), organic matter (bioturbation) and drift (lateral surface processes). However, in more recent years this original dynamic approach has been gradually submerged by zonalism (Gibbs 1980), so that little now remains of the original foundations. Attempts at a zonalistic approach also occurred in New Guinea following extensive land resource surveys in the late 1950s and 1960s that showed the common occurrence of a humic brown clay on a variety of lithologies. This was claimed to be the zonal soil of the humid tropical mountains (Haantjens & Rutherford 1964). Subsequent work, however, established that all these soils were formed on tephra and hence there was no basis for a zonal interpretation (Wood 1987, Humphreys 1991).

Comments regarding the uniqueness of the soils of these compressive margins are perhaps nowhere more common than on the west coast of the USA, where the various USDA schemes of classification have had to deal with this problem. It is doubtful if this has ever been more colourfully expressed than by Geoffrey Milne, who as a result of his 1938 visit wrote (Milne 1939: 35).

The pedological papacy in Washington lays down a body of canon law for its priesthood, but speculative thought is rife even amongst the college of cardinals. In Indiana I was witness to one of its inquisitors in dalliance with advanced doctrines and in Berkeley, California, it positively had a prophet on its payroll. Among the minor clergy heresy is freely talked and on the Pacific seaboard there is a dissident church whose unorthodoxy almost amounts to a schism.

It is unfortunate that the schismatic tendencies, so evident in the 1930s, have been suppressed under the overwhelming influence of zonalism, which continues to the present day. Until space is made for a complete reassessment of pedology, in which the Earth science basis of the subject is fully acknowledged, as has been attempted in this book, there is no possibility of resolving these major problems raised by the zonal approach.

APPENDIX 1

Silicate structures

There are two major factors involved in silicate structures: ionic size and valency. In terms of size it is possible to arrange just four oxygens symmetrically around a silicon in the form of a tetrahedron (Fig. A1.1). However, with silicon being a quadrivalent cation and oxygen a divalent anion, every silica tetrahedron must have a negative charge of 4. The structure of silicate minerals is best explained by the means adopted to balance this negative charge.

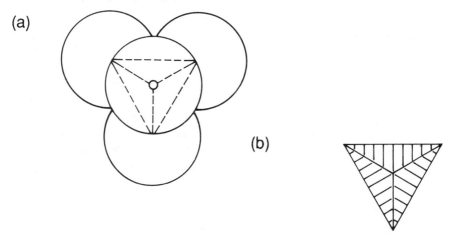

Figure A1.1 The silicon–oxygen tetrahedron: (a) three-dimensional representation; (b) conventional representation.

The most effective way of doing this is by linking tetrahedra (Fig. A1.2) to produce a single chain. Presuming such a chain is of infinite length, any single tetrahedron would share two oxygens with neighbouring tetrahedra and two would remain unshared (Fig. A1.2b), which means that the negative charge on any one tetrahedron as a result of this simple linking would be reduced from 4 to 2.

When two single chains are cross-linked (Fig. A1.3) two types of tetrahedron can be distinguished: external (Fig. A1.3b) and internal (Fig. A1.3c). The external type is linked to two adjoining tetrahedra just as in the case of the single chains, so that for such tetrahedra the excess negative charge remains 2. However, the internal type shares three oxygens with neighbouring tetrahedra, so that its total negative charge becomes 5, leaving an excess of only 1. In this double chain structure there are equal numbers of these two types of tetrahedra so that overall the excess negative charge is 1½ compared to 2 for the single chain and 4 for isolated tetrahedra.

169

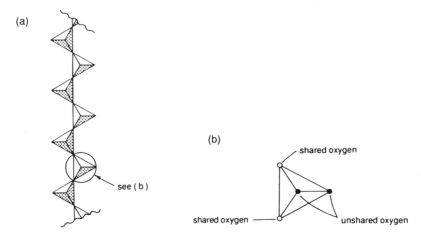

Figure A1.2 Tetrahedral linkage, single chain: (a) standard linkage; (b) details.

The natural progression of increased tetrahedral linkages is to where three oxygens in any one tetrahedron are shared, giving rise to a sheet structure (Fig. A1.4), where all tetrahedra are the same as the internal tetrahedra of the double chain (compare Figs A1.3c and A1.4b) and the charge on any one tetrahedron is reduced to 1.

If the last unshared oxygen is now linked to a neighbouring tetrahedron, a three-dimensional network results, where the four shared oxygens of any tetrahedron give rise to a negative charge of 4, which is balanced by the four positive charges of the silicon. Thus, the linking of silicate tetrahedra by the sharing of all four oxygens gives rise to an electrically neutral silicate framework.

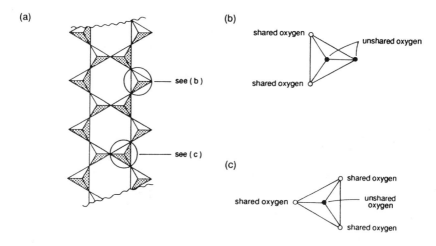

Figure A1.3 Tetrahedral linkage, double chain: (a) standard linkage; (b) details of external tetrahedra; (c) details of internal tetrahedra.

170

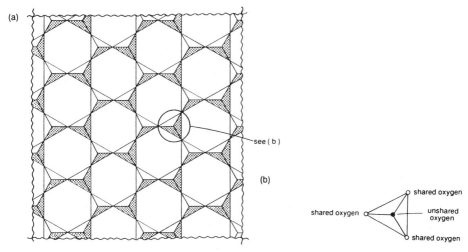

Figure A1.4 Tetrahedral linkage, sheet: (a) standard linkage; (b) details.

It is only where a three-dimensionally linked network (a tektosilicate) is developed that a neutral silicate framework occurs. In all the other cases involving sheets (phyllosilicates), chains (inosilicates) or isolated tetrahedra (orthosilicates) there is a residual negative charge that needs to be balanced by the addition of cations. Some idea of the relative amount of negative charge involved in each of these silicate structures can be gained by making a standard volume of each of them the basis of a comparison. It has been established that in terms of volume any silicate mineral can be regarded as a packing together of oxygen anions, with the cations merely occupying the packing voids, so that if 100 oxygens or 25 tetrahedra are taken from each framework type a standard volume is being considered. In the case of orthosilicates each tetrahedron carries a negative charge of 4 and therefore 25 would have a negative charge of 100. On the same basis single chain inosilicates would have a charge of 50 (2×25), double chain inosilicates $37\frac{1}{2}$ ($1\frac{1}{2} \times 25$), phyllosilicates 25 (1×25) and tektosilicates 0.

Account then needs to be taken of the kind of cation used to balance the overall negative charge, as not all of them are equally suitable for each of the framework types. A major determinant is size (Fig. A1.5). In general, the greater the valency the smaller the size. Thus, univalent potassium is nearly the same size as oxygen, whereas trivalent aluminium is much smaller, approaching quadrivalent silicon is size. Whether or not a particular cation is acceptable within a given framework is dependent on the size of the gaps within it. Thus, potassium would require large gaps, whereas magnesium and iron could make do with much smaller holes. In general it has been found that the more tetrahedral linkages there are in any particular framework the larger the gaps, or, putting it another way, the greater the number of Si–O–Si bonds the more close packing is resisted. Thus, orthosilicates are tightly packed, inosilicates less so, as are phyllosilicates, whereas tektosilicates generally, with the notable exception of quartz, have quite open structures, and this leads to a concentration of the smaller cations, such as magnesium and iron, in orthosilicates and the large cations, potassium, calcium and sodium, in the tektosilicates.

171

The tendency to cation segregation is reinforced by the type of bonds developed. In ionic bonds electrons are transferred to produce a negative and a positive ion, whereas in covalent bonds the electrons are shared equally. However, in silicates it is not an "either/or" affair, but one in which the bonds partake of both ionic and covalent characteristics, which is expressed in terms of electronegativity; that is, the power of an atom in a molecule to attract electrons. When two atoms of the same electronegativity are joined by a bond it will be covalent, since both atoms attract the bonding electrons to the same extent. If, however, one of the atoms has a higher electronegativity than the other, it will draw the shared electrons nearer to itself, giving an ionic component to the bond. The most important bond of the silicate framework is of course that between oxygen and silicon, and it has been found that the greater the number of such bonds in particular frameworks the greater is the electronegativity of oxygen. In other words the electronegativity of oxygen increases through the sequence orthosilicate, inosilicate, phyllosilicate, tektosilicate, and those cations with low electronegativity values (Table A1.1), such as sodium and potassium, will be preferentially included in the tektosilicates.

Table A1.1 Electronegativities.

Na	K	Mg	Ca	Fe	Al	Si	O
0.93	0.82	1.31	1.00	1.83	1.61	1.90	3.44

In addition, as long as cations do not differ too much in size or in valency, they can substitute for one another in silicates. Thus, sodium and calcium can replace one another completely, as can magnesium and ferrous ion (Fig. A1.5). Aluminium, because of its intermediate size and valency, is capable of substituting not only for magnesium and iron, but also in certain circumstances for silicon. Aluminium has a lower electronegativity (1.6) than silicon (1.9) (Table A1.1). The electronegativity of the oxygen in silicates increases from orthosilicates to tektosilicates, so that in the same direction there is an increasing tendency for aluminium to be preferred to silicon in the tetrahedral position. Thus, in some tektosilicates, up to one-quarter of the tetrahedral silicon can be replaced by aluminium, whereas no such substitution can occur in the orthosilicates. There are yet further consequences of this aluminium substitution in tektosilicates, for aluminium is only trivalent whereas silicon is quad-

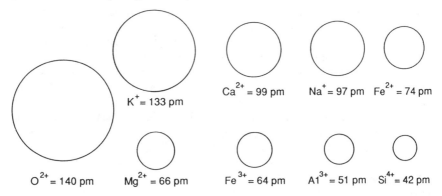

$K^+ = 133$ pm $Ca^{2+} = 99$ pm $Na^+ = 97$ pm $Fe^{2+} = 74$ pm

$O^{2+} = 140$ pm $Mg^{2+} = 66$ pm $Fe^{3+} = 64$ pm $Al^{3+} = 51$ pm $Si^{4+} = 42$ pm

Figure A1.5   Relative size of major ions (in picometres, 10^{-12} m).

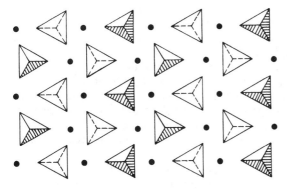

Figure A1.6 Orthosilicate, olivine.

● Fe + Mg

rivalent, which leads to the development of a negative charge. This is balanced by the inclusion of cations of low electronegativity and large size, such as potassium, calcium and sodium, which gives rise to the feldspars. To show how these different factors are combined, an example of a mineral from each of the main framework groups will be discussed.

Olivine is a typical orthosilicate (Fig. A1.6) and its individual tetrahedra alternately point in opposite directions to achieve close packing, for there are no inter-tetrahedral links to preserve an open structure. The negative charge of the individual tetrahedra is balanced by divalent magnesium and iron, which are small enough to fit into the packing voids, the only gaps that occur within the orthosilicates.

Pyroxenes are typical single chain inosilicates (Fig. A1.7) that when viewed end-on can be represented as a double tetrahedral unit (Fig. A1.7b). In Figure A1.7c a view of a pyroxene is given in which the single chains run at right angles to the plane of the page and where in any one row of chains the positions of apical and basal oxygens alternate. The position of the non-framework cations between the rows of the chains is indicated. They can be ferrous and ferric iron, calcium, magnesium and aluminium. In addition aluminium can replace some of the tetrahedral silicon.

Amphiboles are double chain inosilicates (Fig. A1.8). The change to note from the pyroxenes is that when viewed end-on four tetrahedral units are involved (Fig. A1.8b). In Figure A1.8c these double chain units are viewed as running at right angles to the page and, as in the pyroxenes, in any one row of chains the positions of apical and basal oxygens alternate. The position of non-framework cations is also indicated, and in this there is a significant change from the pyroxenes: for where apical oxygens of the chain tetrahedra are pointing towards one another, a wide range of cations occur, such as iron, calcium, magnesium and aluminium, but where the tetrahedral basal surfaces face one another, only the large potassium ion occurs.

Micas have a sheet structure, which can be derived from that of the amphiboles by changing the tetrahedral chains in any one row so that they all face in the same direction and in the next row they all face in the opposite direction. This creates pairs of sheets that are alternately facing and opposed to one another (Fig. A1.9). At the same time this causes all the apical spaces (i) to lie between facing sheets, whereas the basal space (ii) comes to lie between opposing sheets (Figs A1.8, A1.9). The interlayer cations of space (ii) consist only of potassium, which is included within the mica structure to balance the loss of charge resulting from the substitution of some of the tetrahedrally coordinated silicon by trivalent aluminium. The more varied cations of space (i) are all coordinated with six oxygens or hydroxyls, which are octahedrally disposed around them, so that together they can be considered to form an octahedral layer enclosed within two tetrahedral layers. Such 2:1 phyllosilicates can be further classified according to the nature of the octahedrally coordinated cations. In any given phyllosilicate unit cell there are three octahedral sites, and if all three are occupied by divalent cations, e.g. Mg^{2+}, Fe^{2+}, a **trioctahedral mica** such as **biotite** is formed. If, however, only trivalent cations occur (e.g. Al^{3+}, Fe^{3+}) only two out of the three sites are occupied and a **dioctahedral mica** such as **muscovite** is formed.

Feldspars are tektosilicates with a typical three-dimensional framework. The high degree of framework linkage and the resulting high electronegativity values of the oxygen have led to the replacement of one quarter of the tetrahedral silicon by aluminium. The consequent charge deficit has been balanced by the admission of potassium, sodium and calcium into the many large holes in the framework.

Quartz, as mentioned above, is exceptional in being a tightly packed tektosilicate consisting only of silicon and oxygen, which makes it highly inert under typical near-surface conditions.

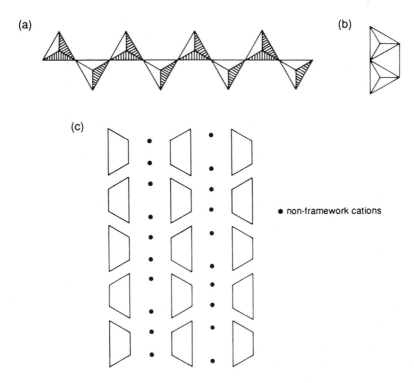

Figure A1.7 Inosilicate, pyroxene: (a) basic chain; (b) end-on view of chain; (c) packing of chains, end-on view.

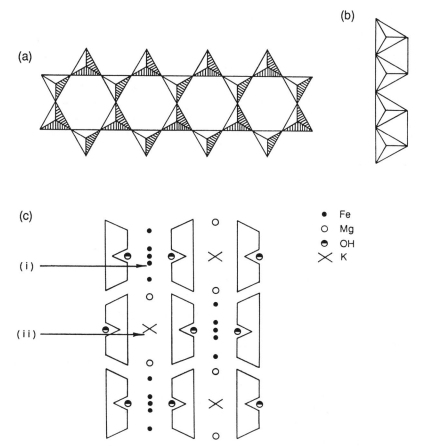

Figure A1.8 Inosilicate, amphibole: (a) basic chain; (b) end-on view of chain; (c) packing of chains, end-on view.

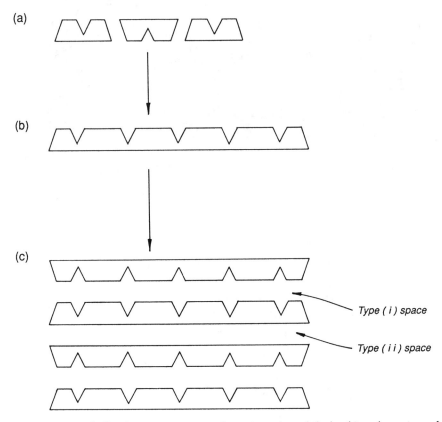

Figure A1.9 Phyllosilicate, mica: (a) end-on view of amphibole; (b) end-on view of mica; (c) packing of mica sheets.

APPENDIX 2

Soil fabric and consistence

Soil fabric assessment requires an understanding of the three-dimensional relationships of both the solid particles and the complementary voids that go to make up soil material in its natural state (Humphreys 1985; Fig. A2.1).

In some soils this three-dimensional relationship is unstable; that is there is a lack of coherency and the material behaves as if it were single-grained. A simple test for coherency, when the soil is in a dry state, is to remove a small amount on a penknife; if it retains its shape it is coherent, if it collapses and runs off the knife it is incoherent.

Coherent materials are characterized by having a stable relationship between solid particles and voids, which can be referred to as the matrix fabric. Given the normal processes of epimorphism acting on the usual types of continental rocks, there is a general tendency for the solid particles produced to be either sand size or clay size (see Part I). Together they form coherent matrix fabrics. If sand grains are sufficiently abundant that they are in contact with one another, a grain-support fabric (GSF) develops; but if there are sufficient clay particles present to surround the sand grains completely, a clay or rather plasma support fabric (PSF) is formed. In terms of field texture grades (see Appendix 3) these grain support fabrics are loamy sands and sandy loams, whereas plasma support fabrics are usually clay loams or heavier. Between sandy loams and clay loams either fabric type occurs, and it is for this reason that texture grade should not be used to determine the type of fabric. The degree of closeness of packing of the individual grains enables porous and dense

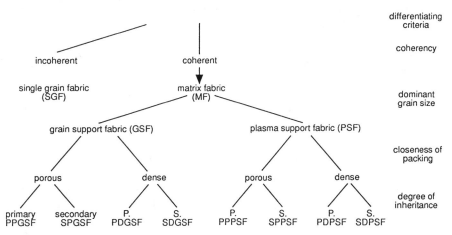

Figure A2.1   Soil fabric (after Humphreys 1985).

177

endpoints to be differentiated in both grain support and plasma support fabrics.

A fourth and final differentiating characteristic with regard to matrix fabrics is the degree to which fabrics are inherited from bedrock or are newly formed. At one extreme a bedrock may be subject to considerable mineralogical change during epimorphism, but the same three-dimensional relationship between the grains persists; that is the fabric but not the mineralogy is inherited from the bedrock and can be regarded as primary. At the other extreme primary fabrics are affected by processes such as mesofaunal activity and rainwash that disrupt existing fabrics and the resulting material has a new or secondary fabric.

Superimposed on the matrix fabric can be coarser void patterns, of which the most important are planar voids, which if abundant delineate peds and also tubules resulting from faunal activity.

From the model of soil formation developed in this book it is seen that there is a fundamental division between the subsoil and the topsoil, for the former has been developed *in situ* (i.e. it is saprolitic) and the latter as a result of movement induced by the mesofauna and rainwash. Such subsoils are primary with a tendency towards dense, plasma support fabrics in the upper part of the saprolite, whereas topsoils are secondary and have mostly grain support fabrics.

Of the coarser void patterns, planar voids are a characteristic feature of subsoils, inherited from bedrock jointing, bedding and sheeting. If such planar voids are sufficiently dense they form a three-dimensional network and hence define peds, which can be further characterized in terms of shape and size. In these circumstances the pattern of planar voids is a primary fabric feature and not secondary as is implied by many soil descriptions. On the other hand tubules, the other characteristic type of coarse void, resulting from faunal activity are for the most part confined to topsoils and are of course secondary.

Within the topsoil it is also possible to differentiate between the very near surface, more recently affected by rainwash, where single-grain fabric is much more common, compared to the immediately underlying topsoil, where porous grain support fabric is more likely because of the dominance of mesofaunal activity. Such observational investigation of soil fabric can be supplemented by testing soil materials in the field both in the natural state, when consistence is determined, and at field capacity, or the sticky point, for field texture grading.

Table A2.1 Soil consistence (after Butler 1955).

brittle	(–2)	Disintegration into finer complex entities or single-grain material almost complete
	(–1)	Some disintegration in which there is a definite reduction in the average size of the entities
	(0)	No definite change beyond possibility of a slight rounding of the entities
	(1)	Coalescence has occurred to produce a considerable number of rod- or ball-shaped entities
plastic	(2)	Almost all the material has coalesced into one mass

Consistence (Butler 1955; Table A2.1) is a measure of how soil material behaves when it is manipulated in its natural state. The test is performed on a 3 cm cube of soil material, which has been removed from the soil mass with as little disturbance as possible. Pressure is exerted on this material by squeezing between the palms of the hand until disruption occurs. The material is then subjected to shearing stress by slowly rubbing the palms of the hands together at the same pressure as that required to cause the initial disruption. Observations are made of the pressure required for disruption, the nature of the fragments produced by the disruption (i.e. to what extent they are pedal) and how these fragments alter with the application of shearing stress, thereby determining the degree of brittleness or plasticity. These determinations are considerably affected by the moisture status of the soil: the degree of change in reaction with change in moisture status can vary from material to material. Therefore, it is desirable for determinations to be made in both dry and moist states. It is, of course, difficult to apply this consistence test to sandy materials and to finer-textured materials in the dry state. Its use is, therefore, restricted in comparison with field texture grading, but it can be of great value when it is applied to complement and amplify observations on fabric. Since most grain support fabrics will be brittle and plasma support fabrics plastic, consistence determinations can be an important confirmatory device.

APPENDIX 3

Field texture grading

Field texture grading is the assessment of soil material behaviour in an homogenized state at the sticky point, which gives particular insight into the fabric of soil material as well as into particle-size distribution. This is accomplished by taking enough soil material to fit into the palm of the hand in a state where it can readily absorb water. Water is added a little at a time and kneaded in, until a uniform ball of material (a bolus) is obtained, which just fails to stick to the fingers. When just at the sticky point the bolus tends to clean the hand as it is worked, but if the sticky point is exceeded the hand will become dirty very quickly, with soil material tending to flow up the arm, and more soil needs to be added. In the normal course of working a bolus, water is absorbed and so it falls below the sticky point; therefore, it may be necessary to add a little water from time to time, to maintain it at the correct point for testing.

The field texture grade is assessed mainly by manipulating the bolus in the hand:
- by overall compression
- by compression between the thumb and fingers
- by shearing between thumb and forefinger (ribboning)
- by rubbing a thin film of soil between thumb and fingers.

Using these methods and taking care to maintain the bolus at the sticky point, it is possible to determine whether the material being tested belongs to one of the following categories:
- The material is a **sand** if the bolus is unstable and disintegrates on minimal working.
- The material is a **loam** if the bolus cracks, but remains coherent on working, a rather fuzzy fingerprint is taken by the bolus and at best unstable ribbons are formed.
- The material is **clay** if the bolus is coherent and requires a noticeable effort to work, it does not crack on working, produces stable ribbons and takes a clear fingerprint.

Within the sands (*sensu lato*) there is a division between the virtually pure **sands** (*sensu stricto*) and **loamy sands** associated with minimal amounts of fines, which gives slight coherence to the bolus and stains the hands to a certain extent.

Loams (*sensu lato*) (Fig. A3.1a) are rather more complex for they change their character depending on admixture with either sand or clays. **Loams** (*sensu stricto*) have a spongy or slightly elastic bounce when compressed in the hand. This characteristic disappears when more sand is present, forming a **sandy loam**, or when more clay is present, forming a **clay loam**, or when both are present and a **sandy clay loam** results.

Clays are more complex for they can vary in three main ways (Fig. A3.1b). In the

180

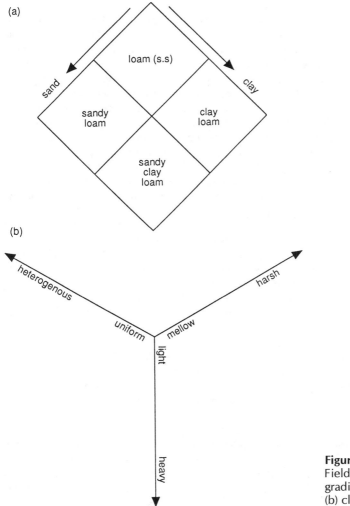

(a)

loam (s.s)

sand

clay

sandy
loam

clay
loam

sandy
clay
loam

(b)

heterogenous

harsh

uniform

mellow

light

heavy

Figure A3.1
Field texture
grading: (a) loams;
(b) clays.

first clays may be uniform or heterogenous depending on the degree to which sand is present. If there is sufficient sand to alter the behaviour of the bolus a **sandy clay** occurs. It is also possible to recognize a continuum between **mellow** and **harsh clays**. The harsh clays are difficult to get into the bolus form, for generally they are very hard when dry and it is necessary to crush them with a hammer, before working the material into a bolus. Even then further difficulties can arise if the crushing has not been fine enough when water is first added, for it does not readily penetrate into even small particles, as clay deflocculates on their outside to form a seal. The end result of this is rather like a handful of ball-bearings, with a mass of deflocculated clay, which runs along the arm. Even if all these difficulties are overcome there is a great tendency for the bolus to be sticky and it requires much greater effort to work than other clays, frequently causing the hand to ache in the process. Many of these harsh clays are 2:1 swelling physils that can be detected in the field by smearing some of the bolus over the hand or arm. On drying the smear contracts and cracks and this

181

causes a tingling sensation. This technique is particularly useful when the clays are fully expanded and no cracks are apparent. In contrast the mellow clays crumble down very easily to a state where water is absorbed with ease to produce a bolus with a certain amount of body, but one that is relatively easy to work.

A third clay characteristic is that of workability, which varies from **light** to **medium** to **heavy**. In addition, it is possible to distinguish an important subset within this sequence and this is for materials that become increasingly difficult to work with time. Such materials are **subplastic** and is the result of iron oxides keeping the clays in a highly flocculated state. In initially forming a bolus a great deal of this flocculation persists, so that larger particles than the ultimate very fine oxides and clay minerals are being dealt with. However, with continued working of the bolus more and more of these compound flocculated particles break down into finer particles, so that with increased time of bolus manipulation the clay content appears to increase, or at least the bolus requires more and more effort to manipulate. The easiest way to detect subplasticity is to make two boluses from the same sample and to stop working one of them immediately a uniform bolus at the sticky point is achieved. Continue working the other bolus for 2–3 minutes, taking particular care to keep it at the sticky point, for a great deal of water is absorbed during this process of particle breakdown. Then work both boluses together, one in each hand, when it is possible to detect differences in their ease of manipulation. In a detailed description all clays should be described in terms of these three characteristics.

In addition it is possible to use other qualifying terms. Thus, considerations of sandiness can be taken one step further and a differentiation made between coarse and fine sand. If when rubbing soil material between the thumb and fingers it is possible to detect individual sand grains, a **coarse** sand is being dealt with. However, if no individual grains are detected, but only a general feeling of roughness, this a **fine** sand. These terms can be used as additional qualifiers anywhere the term sandy is used as well as throughout the sands.

Another qualifier within the loams and clays is the term **silty**, which becomes necessary when there are sufficient phyllosilicate fragments of an illitic type in the silt and very fine sand range to give a silky feel to loams when ribboned. In clays, however, this silky feel is not readily apparent and it is necessary to bite into a small piece of the bolus. If silt-size particles are present they set the teeth on edge whereas pure clays have no effect.

The silky feel of silty material is exactly the same as that attributable to the presence of organic matter, but there is little possibility of confusing the two, for the dark colour of the organic matter and the way it stains the hand should be sufficient to distinguish it. In addition if there is a considerable amount of organic matter, there is generally some difficulty in getting water penetration in the dry state because of its hydrophobic properties. Another characteristic of the presence of organic matter in soil material is that it tends to lighten the field texture grading; thus it could change a clay into a clay loam. In all such cases the qualifier "organic" should be used: **organic clay loam**.

References

Abaturov, B. D. 1972. The role of burrowing animals in the transport of mineral substances in the soil. *Pedobiologia* **12**, 261–6.

Adamson, D., P. M. Selkirk, P. B. Mitchell 1983. The role of fire and lyrebirds in the sandstone landscape of the Sydney Basin. In *Aspects of Australian sandstone landscapes*, R. W. Young & G. C. Nanson (eds), 81–93. Australian and New Zealand Geomorphology Research Group Special Publication 1. Wollongong: University of Wollongong.

Aloni, K. & J. Soyer 1987. Cycle des matériaux de construction des termitières d'humivores en savane au Shaba méridional (Zaïre). *Revue Zoologie Africaine* **101**, 329–58.

Anand, R. R., R. J. Gilkes, T. M. Armitage, J. W. Hillyer 1985. Feldspar weathering in lateritic saprolite. *Clays and Clay Minerals* **33**, 31–43.

Andrews, E. A. 1925. Growth of ant mounds. *Psyche* **32**, 75–87.

Askew, G. P., D. J. Moffatt, R. F. Montgomery, P. L. Searl 1970. Soil landscapes in north-eastern Mato Grosso. *Geographical Journal* **136**, 211–27.

Atkinson, G. 1984. Erosion damage following bushfires. *Journal of the Soil Conservation Service of NSW* **40**, 4–9.

Bagine, R. K. N. 1984. Soil translocation by termites of the genus *Odontotermes* (Holmgren) (Isoptera: Macrotermitinae) in an arid area of northern Kenya. *Oecologia* **64**, 263–6.

Bancalari, M. A. E. & D. A. Perry 1987. Distribution and increment of biomass in adjacent young Douglas-fir stands with different early growth rates. *Canadian Journal of Forest Research* **17**, 722–30.

Barley, K. P. 1959. Earthworms and soil fertility, IV: the influence of earthworms on the physical properties of a red brown earth. *Australian Journal of Agricultural Research* **10**, 371–6.

Baxter, F. P. & F. D. Hole 1967. Ant (*Formica cinerea*) pedoturbation in a prairie soil. *Soil Science Society of America, Proceedings* **31**, 425–8.

Beckmann, G. G., C. H. Thompson, G. D. Hubble 1974. Genesis of red and black soils on basalt on the Darling Downs, Queensland, Australia. *Journal of Soil Science* **25**, 265–81.

Berner, R. A. & G. R. Holdren 1977. Mechanism of feldspar weathering, I: some observational evidence. *Geology* **5**, 369–72.

— 1979. Mechanism of feldspar weathering, II: observations of feldspars from soils. *Geochimica et Cosmochimica Acta* **43**, 1173–86.

Berner, R. A., E. L. Sjöberg, M. A. Velbel, M. D. Krom 1980. Dissolution of pyroxenes and amphiboles during weathering. *Science* **207**, 1205–6.

Berry, L. & B. P. Ruxton 1959. Notes on weathering zones and soils on granitic rocks in two tropical regions. *Journal of Soil Science* **10**, 54–63.

Bishop, N. G. & J. L. Culbertson 1976. Decline of prairie dog towns in southwestern North Dakota. *Journal of Range Management* **29**, 217–20.

Bishop, P. M., P. B. Mitchell, T. R. Paton 1980. The formation of duplex soils on hillslopes in the Sydney Basin, Australia. *Geoderma* **23**, 175–89.

Black, T. A. & D. R. Montgomery 1991. Sediment transport by burrowing mammals, Marin County, California. *Earth Surface Processes and Landforms* **16**, 163–72.

Blake, D. H. 1971. Geology and geomorphology of the Morehead–Kiunga area. In *Land resources of the Morehead–Kiunga area, Papua New Guinea*, part V [Land Resource Series 29]. Melbourne: CSIRO.

Blake, D. H. & C. D. Ollier 1971. Geomorphological evidence of Quaternary tectonics in southwestern Papua. *Revue Géomorphologie Dynamique* **19**, 28–32.

Bleeker, P. 1971. Soils of the Morehead–Kiunga area. In *Land resources of the Morehead–Kiunga area, Papua New Guinea*, part VI [Land Resource Series 29]. Melbourne: CSIRO.

— 1983. *Soils of Papua New Guinea*. Canberra: Australian National University Press.

Blevins, R. L., N. Holowaychuk, L. P. Wilding 1970. Micromorphology of the soil fabric at the tree root/soil interface. *Soil Science Society of America, Proceedings* **34**, 460–4.

Blong, R. J. 1982. *The time of darkness*. Canberra: Australian National University Press.

Blong, R. J., S. J. Riley, P. J. Crozier 1982. Sediment yield from runoff plots following a bushfire near Narrabeen Lagoon, NSW. *Search* **13**, 36–8.

Bloomfield, C. 1953. A study of podzolization, part 2. *Journal of Soil Science* **4**, 17–23.

— 1954. A study of podzolization, parts 3, 4 and 5. *Journal of Soil Science* **5**, 39–56.

Bond, R. D. & J. R. Harris 1964. The influence of the microflora on the physical properties of soils, I: Effects associated with filamentous algae and fungi. *Australian Journal of Soil Research* **2**, 123–31.

Bouillon, A. 1970. Termites of the Ethiopian region. In *Biology of termites*, vol. 2, K. Krishna & F. M. Wessner (eds), 153–280. London: Academic Press.

Bowler, J. M. 1973. Clay dunes: their occurrence, formation and environmental significance. *Earth Science Review* **9**, 315–38.

Branner, J. C. 1896. Decomposition of rocks in Brazil. *Geological Society of America, Bulletin* **7**, 255–314.

— 1900. Ants as geological agents in the tropics. *Journal of Geology* **8**, 151–3.

— 1910. Geological work of ants in tropical America. *Geological Society of America, Bulletin* **21**, 449–96.

Brewer, R. 1955. *Mineralogical examination of a yellow podzolic soil formed on granodiorite* [Soil Publication 5]. Melbourne: CSIRO.

— 1964. *Fabric and mineral analysis of soils*. New York: John Wiley.

— 1968. Clay illuviation as a factor in particle size differentiation in soil profiles. *Transactions of the 9th International Congress of Soil Science, Adelaide*, vol. 4, 489–99. Sydney: International Society of Soil Science.

Brewer, R. & E. Bettany 1973. Further evidence concerning the origin of the Western Australian sand plains. *Geological Society of Australia, Journal* **19**, 533–41.

Bricker, O. P., A. E. Godfrey, E. T. Cleaves 1968. Mineral water interaction during the chemical weathering of silicates. *American Chemical Society, Advances in Chemistry* **73**, 128–42.

Bridges, E. M. 1978. *World soils*, 2nd edn. Cambridge: Cambridge University Press.

Briese, D. T. 1982. The effects of ants on the soil of a semi-arid salt bushland habitat. *Insectes Sociaux* **29**, 375–82.

Brooks, R. R. 1987. *Serpentine and its vegetation: a multidisciplinary approach*. London: Croom Helm.

Brown, G. & I. Stephen 1959. A structural study of iddingsite from NSW, Australia. *American Mineralogist* **44**, 251–60.

Bryson, H. R. 1933. The amount of soil brought by insects to the surface of a watered and an unwatered plot. *Kansas Entomological Society, Journal* **6**, 81–90.

Buchanan, R. A. 1980. The Lambert Peninsula, Ku-ring-gai Chase National Park. Physiography and the distribution of podzols, shrublands and swamps, with details of swamp vegetation and sediments. *Linnean Society of NSW, Proceedings* **104**, 73–94.

Buchanan, R. A. & G. S. Humphreys 1980. The vegetation on two podzols on the Hornsby Plateau, Sydney. *Linnean Society of NSW, Proceedings* **104**, 50–71.

Buechner, H. K. 1942. Interrelationships between the pocket gopher and land use. *Journal of Mammalogy* **23**, 346–8.

Burns, S. F. & P. J. Tonkin 1987. Erosion and sediment transport by windblow in a mountain beech forest, New Zealand. In *Erosion and sedimentation in the Pacific Rim*, R. C. Beschta, T. Blinn, G. E. Grant, G. C. Ice, F. J. Swanson (eds), 269–78. Publication 165, International Association of Hydrological Sciences, Wallingford, England.

Butler, B. E. 1955. A system for the description of soil structure and consistence in the field. *Australian Institute of Agricultural Science, Journal* **21**, 239–49.

— 1956. Parna – an aeolian clay. *Australian Journal of Science* **18**, 145–51.

— 1959. *Periodic phenomena in landscapes as a basis for soil studies* [Soil Publication 14].Melbourne: CSIRO.

Butuzova, O. V. 1962. Role of the root system of trees in the formation of microrelief. *Soviet Soil Science* **4**, 364–72.

Butzer, K. W. 1978. Climate patterns in an unglaciated continent. *Geographical Magazine* **51**, 201–8.

Cady, J. G. 1965. Petrographic microscope techniques. In *Methods of soil analysis*, J. L. White, L. E. Ensminger, F. E. Clark (eds), 604–31. Madison, Wisconsin: American Society of Agronomy.

Carey, S. W. 1954. The rheid concept in geotectonics. *Geological Society of Australia, Journal* **1**, 67–117.

Castellanos, J., M. Maass, J. Kummerow 1991. Root biomass in a dry deciduous tropical forest in Mexico. *Plant and Soil* **131**, 225–8.

Catt, J. A. 1986. *Soils and Quaternary geology*. Oxford: Oxford University Press.

Cavelier, J. 1992. Fine-root biomass and soil properties in a semi-deciduous and lower montane rainforest in Panama. *Plant and Soil* **142**, 187–201.

Charter, C. F. 1949a. The detailed reconnaissance soil survey of the cocoa country of the Gold Coast. *Proceedings of the Cocoa Conference 1949*, 19–24. London: The Cocoa, Chocolate and Confectionery Alliance.

— 1949b. The characteristics of the principal cocoa soils. *Proceedings of the Cocoa*

Conference 1949, appendix II, 105–12. London: The Cocoa, Chocolate and Confectionary Alliance.

Chartres, C. J. 1982. Pedogenesis of desert loam soil in the Barrier Range, western New South Wales, I: soil parent materials. *Australian Journal of Soil Research* **21**, 1–13.

Chenery, E. M. 1948. Aluminium in the plant world, part 1: general survey in dicotyledons. *Kew Bulletin* **2**, 173–83.

Chittleborough, D. J. & J. M. Oades 1979. The development of a red brown earth, I: a reinterpretation of published data. *Australian Journal of Soil Research* **17**, 371–81.

— 1980a. The development of a red brown earth, II: uniformity of the parent material. *Australian Journal of Soil Research* **18**, 375–82.

— 1980b. The development of a red brown earth, III: the degree of weathering and translocation of the clay. *Australian Journal of Soil Research* **18**, 383–93.

Chittleborough, D. J., P. H. Walker, J. M. Oades 1984a. Textural differentiation in chronosequences from eastern Australia, I: descriptions, chemical properties and micromorphologies of soils. *Geoderma* **32**, 181–202.

— 1984b. Textural differentiation in chronosequences from eastern Australia, II: evidence from particle size distributions. *Geoderma* **32**, 203–26.

— 1984c. Textural differentiation in chronosequences from eastern Australia, III: evidence from elemental chemistry. *Geoderma* **32**, 227–48.

Coleman, D. C. 1976. A review of root production processes and their influence on soil biota in terrestrial ecosystems. In *The role of terrestrial and aquatic organisms in decomposition processes*, J. M. Anderson & A. MacFadyen (eds), 417–34. Oxford: Blackwell Scientific.

Coulson, C. B., R. I. Davies, D. A. Lewis 1960. Polyphenols in plant humus and soils, parts 1 and 2. *Journal of Soil Science* **11**, 20–44.

Coventry, R. J. 1982. The distribution of red, yellow and grey earths in the Torrens Creek area, central north Queensland. *Australian Journal of Soil Research* **20**, 1–14.

Cowan, J. A., G. S. Humphreys, P. B. Mitchell, C. L. Murphy 1985. An assessment of pedoturbation by two species of mound building ants, *Camponotus intrepidus* (Kirby) and *Iridomymex purpureus* (F. Smith). *Australian Journal of Soil Research* **22**, 98–108.

Dare-Edwards, A. J. 1984. Aeolian clay deposits of southeastern Australia: parna or loessic clay. *Institute of British Geographers, Transactions* **9**, 337–44.

Darmody, R. G. & J. E. Foss 1982. Soil landscape relationships in the Piedmont of Maryland. *Soil Science Society of America, Journal* **46**, 588–92.

Darwin, C 1881. *The formation of vegetable mould through the action of worms, with observations on their habits*. London: John Murray.

Dash, M. C. & V. C. Patra 1979. Wormcast production and nitrogen contribution to soil by a tropical earthworm population from a grassland site in Orissa, India. *Revue d'Ecologie Biologie du Sol* **16**, 79–83.

Davies, R. I., C. B. Coulson, D. A. Lewis 1964. Polyphenols in plant humus and soils, parts 3 and 4. *Journal of Soil Science* **15**, 299–318.

Davis, W. M. 1892. The convex profiles of bad land divides. *Science* **20**, 245.

De Bano, L. F., L. D. Mann, D. A. Hamilton 1970. Translocation of hydrophobic substances into soil by burning organic litter. *Soil Science Society of America,*

Proceedings **34**, 130–3.

De Meester, T. 1970. *Soils of the Great Konya Basin, Turkiye.* Agricultural Research Report 740, Centre for Agricultural Publication and Documentation, Wageningen.

Denny, C. S. & J. C. Goodlett 1956. *Microrelief resulting from fallen trees.* USGS Professional Paper 288, 59–68.

De Vore, G. W. 1959. The surface chemistry of feldspars as an influence on their decomposition products. *Clays and Clay Minerals* **2**, 26–41.

Dietrich, W. E., T. Dunne, N. F. Humphrey, L. M. Reid 1982. Construction of sediment budgets for drainage basins. In *Workshop on sediment budgets and routing in forested drainage basins* [General Technical Report PNW-141], F. J. Swanson, R. J. Janda, T. Dunne, D. N. Swanston (ed.), 5–23. Washington, DC: US Forest Service.

Drummond, H. 1884–5. On the termite as the tropical analogue of the earthworm. *Royal Society of Edinburgh, Proceedings* **13**, 137–46.

Duchaufour, P. 1982. *Pedology.* London: Allen & Unwin.

Dunin, A. & E. Ganor 1991. Trapping of airborne dust by mosses in the Negev Desert, Israel. *Earth Surface Processes and Landforms* **16**, 153–62.

Dymond, J., P. E. Biscaye, R. W. Rex 1974. Aeolian origin of mica in Hawaiian soils. *Geological Society of America, Bulletin* **85**, 37–40.

Edwards, C. A. & J. R. Lofty 1977. *Biology of earthworms.* London: Chapman & Hall.

Ellison, L. 1946. The pocket gopher in relation to soil erosion on mountain ranges. *Ecology* **27**, 101–14.

Emmett, W. W. 1970. *The hydraulics of overland flow on hillslopes.* USGS Professional Paper 662A.

Eswaran, H. 1979. The alteration of plagioclases and augites under differing pedo-environmental conditions. *Journal of Soil Science* **30**, 547–55.

Eswaran, H. & Y. Y. Heng 1976. The weathering of biotite in a profile on gneiss in Malaysia. *Geoderma* **16**, 9–20.

Eswaran, H. & Wong Chaw Bin 1978. A study of a deep weathering profile on granite in peninsular Malaysia, I: Physicochemical and micromorphological properties. *Soil Science Society of America, Proceedings* **46**, 144–53.

Evans, A. C. 1948. Studies on the relationships between earthworms and soil fertility, II: some effects of earthworms on soil structure. *Annals of Applied Biology* **35**, 1–13.

Evans, A. C. & W. J. McL. Guild 1947. Studies of the relationships between earthworms and soil fertility, I: biological studies in the field. *Annals of Applied Biology* **34**, 307–30.

Fanning, D. S., V. Z. Keremidas, M. A. El-Desoky 1989. Micas. In *Minerals in soil environments* (2nd edn), J. B. Dixon & S. B. Weed (eds), 551–634. Madison: Soil Science Society of America.

FAO–UNESCO 1970–80. *Soil map of the world, 1:5 000 000.* Vols 1–10. Paris: UNESCO.

Fawcett, J. J. 1965. Alteration products of olivine and pyroxene in basalt lavas from the Isle of Mull. *Mineralogical Magazine* **35**, 55–68.

Fieldes, M. 1955. Clay mineralogy of New Zealand soils, part 2: allophane and related mineral colloids. *New Zealand Journal of Science and Technology* **B37**, 336–50.

187

Fieldes, M. & L. D. Swindale 1954. Chemical weathering of silicates in soil formation. *New Zealand Journal of Science and Technology* **B36**, 140–54.

Finlayson, B. L. 1985. Soil creep: a formidable fossil of misconception. In *Geomorphology and soils*, K. S. Richards, R. R. Arnett, S. Ellis (eds), 141–58. London: Allen & Unwin.

Flint, R. F. 1971. *Glacial and Quaternary geology.* New York: John Wiley.

Frei, E. & M. G. Cline 1949. Profile studies of normal soils of New York, II: micromorphological studies of the grey brown podzolic–brown podzolic soil sequence. *Soil Science* **68**, 333–44.

Frith, H. J. (ed.) 1976. *Complete book of Australian birds.* Sydney: Reader's Digest.

Forcella, F. 1977. Ants on a Holocene mudflow in the Coast Range of Oregon. *Soil Survey Horizons* **18**, 3–8.

Gakahu, C. G. & G. W. Cox 1984. The occurrence and origin of mima mound terrain in Kenya. *African Journal of Ecology* **22**, 31–42.

Genelly, R. E. 1965. Ecology of the common mole rat (*Cryptomys hotentotus*) in Rhodesia. *Journal of Mammalogy* **46**, 647–65.

Giardino, J. R. 1974. When elephants destroy a valley. *Geographical Magazine* **47**, 175–81.

Gibbs, H. S. 1980. *New Zealand soils.* Wellington: Oxford University Press.

Gilkes, R. J. 1973. The alteration products of potassium depleted oxybiotite. *Clays and Clay Minerals* **21**, 303–13.

Gilkes, R. J. & A. Suddiprakarn 1979. Biotite alteration in deeply weathered granite, I: morphological, mineralogical and chemical properties. *Clays and Clay Minerals* **27**, 349–60.

Gilkes, R. J., R. C. Young, J. P. Quirk 1972. Oxidation of ferrous iron in biotite. *Nature, Physical Sciences* **236**, 89–91.

Gilkes, R. J., G. Scholz, G. M. Dimmock 1973a. Lateritic deep weathering of granite. *Journal of Soil Science* **24**, 523–36.

— 1973b. Artificial weathering of oxidised biotite. *Soil Science Society of America, Proceedings* **37**, 25–33.

Glover, P. E., E. C. Trump, L. E. D. Wateridge 1964. Termitaria and vegetation patterns on the Loita Plains of Kenya. *Journal of Ecology* **52**, 365–77.

Godfrey, G. K. & P. Crowcroft 1960. *The life of the mole*, Talpa europae. London: Museum Press.

Goldrick, G. 1990. *A study of the formation and erosion of soil on Booberoi regeneration area, western NSW.* Honours thesis, School of Earth Sciences, Macquarie University, Sydney.

Goudie, A. S. 1988. The geomorphological role of termites and earthworms in the tropics. In *Biogeomorphology*, H. A. Viles (ed.), 166–92. Oxford: Basil Blackwell.

Greaves, T. 1962. Studies of foraging galleries and the invasion of living trees by *Coptotermes acinaciformis* and *C. bruneus* (Isoptera). *Australian Journal of Zoology* **10**, 630–51.

Green, P. 1966. Mineralogical and weathering study of a red brown earth formed on granodiorite. *Australian Journal of Soil Research* **4**, 181–97.

Greenslade, P. J. M. 1974. Some relations of the meat ant, *Iridiomymex purpureus* (Hymenoptera: Formicidae) with soil in south Australia. *Soil Biology & Biochemistry* **6**, 7–14.

Greenway, P. 1980. Water balance and urine production in the Australian arid zone crab, *Hothuisana transversa*. *Journal of Experimental Biology* **87**, 237–46.

Greenwood, J. E. G. W. 1957. The development of vegetation patterns in Somaliland Protectorate. *Geographical Journal* **123**, 465–73.

Grinnell, J. 1923. The burrowing rodents of California as agents in soil formation. *Journal of Mammalogy* **4**, 137–48.

Guild, W. J. McL. 1955. Earthworms and soil structure. In *Soil zoology*, D. K. McE. Kevan (ed.), 88–98. London: Butterworth.

Gunn, R. H. 1967. A soil catena on denuded laterite profiles in Queensland. *Australian Journal of Soil Research* **5**, 117–32.

Gupta, S. R., R. Rajvanshi, J. S. Singh 1981. The role of the termite *Odontotermes gurdaspurensis* (Isoptera: Termitidae) in plant decomposition in a tropical grassland. *Pedobiologia* **22**, 254–61.

Haantjens, H. A. & G. K. Rutherford 1964. Soil zonality and parent rock in a very wet tropical mountain region. *Transactions of the 8th International Congress of Soil Science*, vol. 5, 493–500. Bucharest: International Society of Soil Science.

Hallsworth, E. G. & G. G. Beckman 1969. Gilgai in the Quaternary. *Soil Science* **107**, 407–20.

Hallsworth, E. G., G. K. Robertson, F. R. Gibbons 1955. Studies in pedogenesis in New South Wales, VII: gilgai soils. *Journal of Soil Science* **6**, 1–31.

Handley, W. R. C. 1954. *Mull and mor formation in relation to forest soils*. Bulletin 23, UK Forestry Commission, London.

Harker, A. 1950. *Metamorphism*, 3rd edn. London: Methuen.

Hart, D. M. 1988. A fabric contrast soil on dolerite in the Sydney Basin, Australia. *Catena* **15**, 27–37.

— 1992. *The role of plant opal in the Australian environment*. PhD thesis, School of Earth Sciences, Macquarie University.

Hart, D. M., P. P. Hesse, P. B. Mitchell 1985. The inheritance of soil fabric from joints in the parent rock. *Journal of Soil Science* **36**, 367–72.

Hazelhoff, L., P. van Hoff, A. C. Imeson, F. J. P. M. Kwaad 1981. The exposure of forest soil to erosion by earthworms. *Earth Surface Processes and Landforms* **6**, 235–50.

Hemming, C. F. 1965. Vegetation arcs in Somaliland. *Journal of Ecology* **53**, 57–68.

Hesse, P. P. 1985. *Soil stratigraphy of two small tableland valleys*. Honours thesis, School of Earth Sciences, Macquarie University.

Hole, F. D. 1981. Effects of animals on soils. *Geoderma* **25**, 75–112.

Holmes, A. 1965. *Principles of physical geology*, 2nd edn. London: Thomas Nelson.

Holt, J. A., R. J. Coventry, D. F. Sinclair 1980. Some aspects of the biology and pedological significance of mound-building termites in a red and yellow earth landscape near Charters Towers, North Queensland. *Australian Journal of Soil Research* **18**, 97–109.

Huang, W. H. & W. D. Keller 1972. Organic acids as agents of chemical weathering of silicate minerals. *Nature, Physical Sciences* **239**, 149–51.

Hughes, J. C. 1980. Crystallinity of kaolin minerals and their weathering sequence in some soils from Nigeria, Brazil and Colombia. *Geoderma* **24**, 317–25.

Hughes, J. C. & G. Brown 1979. A crystallinity index for soil kaolins and its relation to parent rock, climate and soil maturity. *Journal of Soil Science* **30**, 557–563.

Humphreys, G. S. 1981. The rate of ant mounding and earthworm casting near Sydney, New South Wales. *Search* **12**, 129–31.

— 1984. *The environment and soils of Chimbu Province, Papua New Guinea, with particular reference to soil erosion.* Research Bulletin 35, Department of Primary Industry, Port Moresby.

— 1985. *Bioturbation, rainwash and texture contrast soils.* PhD thesis, School of Earth Sciences, Macquarie University.

— 1989. Earthen structures built by nymphs of the cicada, *Cyclochila australasiae* (Donovan) (Homoptera: Cicadidae). *Australian Entomology Magazine* **16**, 99–108.

— 1991. Soil maps of Papua New Guinea: a review. *Science in New Guinea* **17**, 77–102.

— 1994a. Bioturbation, biofabrics and the biomantle: an example from the Sydney Basin. In *Soil micromorphology: studies in management and genesis*, A. J. Ringrose-Voase & G. S. Humphreys (eds), 421–36. Amsterdam: Elsevier.

— 1994b. Bowl-structures: a composite depositional soil crust. In *Soil micromorphology: studies in management and genesis*, A. J. Ringrose-Voase & G. S. Humphreys (eds), 787–98. Amsterdam: Elsevier.

Humphreys, G. S. & P. B. Mitchell 1983. A preliminary assessment of the role of bioturbation and rainwash on sandstone hillslopes in the Sydney Basin. In *Aspects of Australian sandstone landscapes* [Australian and New Zealand Geomorphology Research Group Special Publication 1], R. W. Young & G. C. Nanson (eds), 65–80. Wollongong: University of Wollongong.

Hunt, C. B. 1966. *Geology of soils.* San Francisco: W. H. Freeman.

Hutchinson, G. E. 1943. The biogeochemistry of aluminium and certain related elements. *Quarterly Review of Biology* **18**, 1–346.

Imeson, A. C. 1976. Some effects of burrowing animals on slope processes in the Luxembourg Ardennes, part 1: the excavation of animal mounds in experimental plots. *Geografiska Annaler* **58A**, 115–25.

— 1977. Splash erosion, animal activity and sediment supply in a small forested Luxembourg catchment. *Earth Surface Processes and Landforms* **2**, 153–160.

Imeson, A. C. & F. J. P. M. Kwaad 1976. Some effects of burrowing animals on slope processes in the Luxembourg Ardennes, part 2: the erosion of animal mounds by splash under forest. *Geografiska Annaler* **58A**, 317–28.

Imeson, A. C. & H. J. M. van Zon 1980. Erosion processes in small forest catchments in Luxembourg. In *Geographical approaches to fluvial processes*, A. F. Pitty (ed.), 93–107. Norwich: Geo Abstracts.

Ingles, L. G. 1952. The ecology of the mountain pocket gopher, *Thomomys monticola. Ecology* **33**, 87–95.

Inoue, K. & P. M. Huang 1986. Influence of citric acid on the natural formation of imogolite. *Nature* **308**, 58–60.

Ireland, H. A., C. F. S. Sharpe, D. H. Eargle 1939. *Principles of gully erosion in the Piedmont of South Carolina.* USDA Technical Bulletin 633.

Isbell, R. F. & G. P. Gillman 1973. Studies on some deep sandy soils in Cape York Peninsula, North Queensland, I: morphological and chemical characteristics. *Australian Journal of Experimental Agriculture and Animal Husbandry* **13**, 81–8.

Isbell, R. F., C. H. Thompson, G. D. Hubble, G. G. Beckmann, T. R. Paton 1967.

Atlas of Australian soils. Explanatory data for sheet 4: Brisbane–Charleville–Rockhampton–Clermont area. Melbourne: Melbourne University Press.

Iskandar, I. K. & J. K. Syers 1972. Metal complex formation by lichen compounds. *Journal of Soil Science* **23**, 255–65.

Jackson, M. L., T. W. M. Levelt, J. K. Syers, R. W. Rex, R. N. Clayton, G. D. Sherman, G. Uehara 1971. Geomorphological relationships of tropospherically derived quartz in the soils of the Hawaiian Islands. *Soil Science Society of America, Proceedings* **35**, 515–25.

Jackson, M. L., S. A. Tyler, A. L. Willis, G. A. Bourbeau, R. P. Pennington 1948. Weathering sequence of clay size minerals in soils and sediments, I: fundamental generalisations. *Journal of Physical and Colloid Chemistry* **52**, 1237–60.

Jacot, A. P. 1936. Soil structure and soil biology. *Ecology* **17**, 359–79.

Johnson, D. L. 1989. Subsurface stone lines, stone zones, artifact manuport layers and biomantles produced by bioturbation via pocket gophers (*Thomomys bottae*). *American Antiquity* **54**, 370–89.

— 1990. Biomantle evolution and the redistribution of earth materials and artifacts. *Soil Science* **149**, 84–102.

Jonca, E. 1972. Water denudation of molehills in mountainous areas. *Acta Theriologica* **17**, 407–12.

Juang, T. C. & G. Uehara 1968. Mica genesis in Hawaiian soils. *Soil Science Society of America, Proceedings* **32**, 31–5.

Kalisz, P. J. & E. L. Stone 1984. Soil mixing by scarab beetles and pocket gophers in north-central Florida. *Soil Science Society of America, Journal* **48**, 169–72.

Keller, W. D. 1957. *The principles of chemical weathering.* Missouri: Lucas.

Keller, W. D. & A. F. Frederickson 1952. Role of plants and colloidal acids in the mechanism of weathering. *American Journal of Science* **250**, 594–608.

Kellogg, C. E. 1949. *Preliminary suggestions for the classification and nomenclature of great soil groups in tropical and equatorial regions.* Technical Communication 46 (76–85), Commonwealth Bureau of Soil Science, Farnham, England.

Kellog, C. E. & F. D. Davol 1949. *An exploratory study of soil groups in the Belgian Congo* [INEAC Series Scientifique 46]. Brussels: Government Printer.

Key, K. H. L. 1959. The ecology and biogeography of Australian grasshoppers and locusts. In *Biogeography and ecology in Australia*, A. Keast, R. L. Crocker, C. S. Christian (eds), 192–210. The Hague: Junk.

Khodashova, K. S. & L. S. Dinesman 1961. Role of small ground squirrels in the formation of complex soil in clay semi-desert of the trans-Volga region. *Soviet Soil Science* **1**, 55–62.

Kirkman, J. H. 1975. Clay mineralogy of some tephra beds of the Rotarua area, North Island, New Zealand. *Clay Minerals* **10**, 437–49.

Komiyama, A., K. Ogino, S. Aksonkoae, S. Sabhasri 1987. Root biomass of a mangrove forest in Southern Thailand, 1: estimation by the trench method and the zonal structure of root biomass. *Journal of Tropical Ecology* **3**, 97–108.

Krishnamoorthy, R. V. 1985. A comparative study of wormcast production by earthworm populations from grassland and woodland near Bangalore, India. *Revue d'Ecologie et de Biologie du Sol* **22**, 209–19.

Krismannsdòttir, H. 1982. Alteration in the IRDP drill hole compared with other drill holes in Iceland. *Journal of Geophysical Research* **87**(B8), 6525–31.

Kubiena, W. L. 1953. *The soils of Europe*. London: Allen & Unwin (Thomas Murby).

— 1956. Rubefizierung und Laterisierung. *Transactions of the 6th International Congress of Soil Science, Bucharest*, vol. 5, 247–9. Paris: International Society of Soil Science.

Kwaad, F. J. P. M. 1977. Measurement of rainsplash erosion and the formation of colluvium beneath deciduous woodland in the Luxembourg Ardennes. *Earth Surface Processes and Landforms* **2**, 161–73.

Lal, R. 1976. No-tillage effects on soil properties under different crops in western Nigeria. *Soil Science Society of America, Journal* **40**, 763–8.

Lanson, B. & D. Champion 1991. The I/S to illite reaction in the late stage of diagenesis. *American Journal of Science* **291**, 473–506.

Lavelle, P. 1978. *Les vermes de terre de la savane de Lamto (Côte d'Ivoire) peuplements, populations et fonctions dans l'ecosysteme* [Publications du Laboratoire de Zoologie 12]. Paris: Ecole Normale Supérieure.

Lee, K. E. 1985. *Earthworms: their ecology and relationships with soils and land use*. Sydney: Academic Press.

Lee, K. E. & T. G. Wood 1968. Preliminary studies of the role of *Nasutitermes exitiosus* (Hill) in the cycling of organic matter in a yellow podzolic soil under dry sclerophyll forest in South Australia. *Transactions of the 9th International Congress of Soil Science, Adelaide*, vol. 2, 11–18. Sydney: International Society of Soil Science.

— 1971a. *Termites and soils*. London: Academic Press.

— 1971b. Physical and chemical effects on soils of some Australian termites and their pedological significance. *Pedobiologia* **11**, 376–409.

Leow, K. S. & F. T. Degge 1981. Some soil characteristics of termite mounds under Guinea savanna climate, Zaria, Kaduna State, Nigeria. *Malaysian Journal of Tropical Geography* **4**, 33–9.

Lepage, M. 1984. Distribution, density and evolution of *Macrotermes bellicosus* nests (Isoptera: Macrotermitinae) in the north-east of the Ivory Coast. *Journal of Animal Ecology* **53**, 107–17.

Lindquist, A. W. 1933. Amounts of dung buried and soil excavated by certain coprini (Scarabaeidae). *Kansas Entomological Society, Journal* **6**, 109–23.

Ljungström, P. O. & A. J. Reinecke 1969. Ecology and natural history of the microchaeltid earthworms of South Africa, 4: studies on influences of earthworms upon the soil and the parasitological questions. *Pedobiologia* **9**, 152–7.

Lobry de Bruyn, L. A. & A. J. Conacher 1990. The role of termites and ants in soil modification: a review. *Australian Journal of Soil Research* **28**, 55–93.

Lockaby, B. G. & J. C. Adams 1985. Pedoturbation of a forest soil by fire ants. *Soil Science Society of America, Proceedings* **49**, 220–3.

Lovering, T. S. 1959. Significance of accumulator plants in rock weathering. *Geological Society of America, Bulletin* **70**, 781–800.

Lutz, H. J. & F. S. Griswold 1939. The influence of the tree roots on soil morphology. *American Journal of Science* **237**, 389–400.

Lyford, W. H. 1963. Importance of ants to brown podzolic soils genesis in New England. *Harvard Forum Paper* **7**, 1–18.

— 1969. The ecology of an elfin forest in Puerto Rico, 7: soil, root and earthworm relationships. *Arnold Arboretum, Journal* **50**, 210–24.

Mabbutt, J. A. (ed.) 1972. *Lands of the Fowlers Gap–Calindary area, New South Wales.* Research Series 4, Fowlers Gap Arid Zone Research Station, University of New South Wales.

— 1977. *Desert land forms.* Canberra: Australian National University Press.

Macedo, J. & R. B. Bryant 1987. Morphology, mineralogy and genesis of a hydrosequence of oxisols in Brazil. *Soil Science Society of America, Journal* **51**, 690–8.

MacFadyen, W. A. 1950a. Soil and vegetation in British Somaliland. *Nature* **165**, 121.

— 1950b. Vegetation patterns in the semi-desert plains of British Somaliland. *Geographical Journal* **116**, 199–211.

Macleod, D. A. 1980. The origin of the red Mediterranean soils in Epirus, Greece. *Journal of Soil Science* **31**, 125–36.

Madge, D. S. 1965. Leaf fall and litter disappearance in a tropical forest. *Pedobiologia* **5**, 273–88.

— 1969. Field and laboratory studies on the activities of two species of tropical earthworms. *Pedobiologia* **9**, 188–214.

Mandel, R. D. & C. J. Sorenson 1982. The role of the western harvester ant (*Pogonomymex occidentalis*) in soil formation. *Soil Science Society of America, Journal* **46**, 785–8.

Mason, B. 1966. *Principles of geochemistry*, 3rd edn. New York: John Wiley.

Mast, M. A. & J. I. Drever 1987. The effect of oxalate on the dissolution rates of oligoclase and tremolite. *Geochemica et Cosmochimica Acta* **51**, 2559–68.

Mayr, E. 1982. *The growth of biological thought.* Cambridge, Massachusetts: Harvard University Press.

Mazurak, A. P. & P. N. Mosher 1968. Detachment of soil particles in simulated rainfall. *Soil Science Society of America, Proceedings* **32**, 716–9.

McCaleb, S. B. 1959. The genesis of red yellow podzolic soils. *Soil Science Society of America, Proceedings* **24**, 164–8.

McCallien, W. J., B. P. Ruxton, B. J. Walton 1960. Mantle rock tectonics: a study in tropical weathering at Accra, Ghana. *Overseas Geological and Mineral Research* **9**, 257–94.

McIlroy, J. C., R. J. Cooper, E. J. Gifford 1981. Inside the burrow of the common wombat, *Vombatus ursinus* (Shaw 1800). *Victorian Naturalist* **98**, 60–5.

McKeague, J. A. 1983. Clay skins and argillic horizons. In *Soil micromorphology*, vol. 2: *soil genesis*, P. Bullock & C. Murphy (eds), 367–87. Berkhamsted, England: A. B. Academic Publications.

McKeague, J. A., R. K. Guertin, F. Page, K. W. G. Valentine 1978. Micromorphological evidence of illuvial clay in horizons designated Bt in the field. *Canadian Journal of Soil Science* **58**, 179–86.

McKeague, J. A., R. K. Guertin, K. W. G. Valentine, J. Belisle, G. A. Bourbeau, A. Howell, W. Michalyna, L. Hopkins, F. Page, L. M. Bresson 1980. Estimating illuvial clay in soils by micromorphology. *Soil Science* **129**, 386–8.

McKeague, J. A., C. Wang, G. J. Ross, C. J. Acton, R. E. Smith, D. W. Anderson, W. W. Pettapiece, T. M. Lord 1981. Evaluation of criteria for argillic horizons (Bt) of soils in Canada. *Geoderma* **25**, 63–74.

McTainsh, G. 1984. The nature and origin of the aeolian mantles of central Northern Nigeria. *Geoderma* **33**, 13–37.

— 1985. Dust processes in Australia and West Africa: a comparison. *Search* **16**, 104–6.

Mehegan, J. M., P. T. Robinson, J. P. Delaney 1982. Secondary mineralization and hydrothermal alteration in the Reydarfjordur drill core, eastern Iceland. *Journal of Geophysical Research* **87**(B8), 6511–24.

Mellanby, K. 1971. *The mole*. London: Collins.

Melton, D. A. 1976. The biology of the aardvark (*Tubulidentata orycteropodiae*). *Mammal Review* **6**, 75–88.

Milne, G. 1935. Some suggested units of classification and mapping, particularly for East African soils. *Soil Research* **4**, 183–98.

— 1936. Normal erosion as a factor in soil profile development. *Nature* **138**, 548–9.

— 1939. *Report on a journey to part of the West Indies and the United States for the study of soils, February to August 1938*. Dar-es-Salaam: Government Printer.

— 1947. A soil reconnaissance journey through parts of Tanganyika Territory, December 1935 to February 1936. *Journal of Ecology* **35**, 192–265.

Mitchell, P. B. 1985. *Some aspects of the role of bioturbation in soil formation in southeastern Australia*. PhD thesis, School of Earth Sciences, Macquarie University.

— 1988. The influences of vegetation, animals and micro-organisms on soil processes. In *Biogeomorphology*, H. A. Viles (ed.), 43–82. Oxford: Basil Blackwell.

Mitchell, P. B. & G. S. Humphreys 1987. Litter dams and microterraces formed on hillslopes subject to rainwash in the Sydney Basin, Australia. *Geoderma* **39**, 331–57.

Moniz, A. C. & S. W. Buol 1982. Formation of an oxisol–ultisol transition in São Paulo, Brazil, I: double water flow model of soil development. *Soil Science Society of America, Journal* **46**, 1228–33.

Moniz, A. C., S. W. Buol, S. B. Weed 1982. Formation of an oxisol–ultisol transition in São Paulo, Brazil, II: lateral dynamics of chemical weathering. *Soil Science Society of America, Journal* **46**, 1234–9.

Morales, C. (ed.) 1979. *Saharan dust: mobilisation, transport, deposition* [SCOPE 14]. Chichester: John Wiley..

Morgan, J. P., J. M. Coleman, S. M. Gagliano 1968. Mudlumps: diapiric structures in Mississippi delta sediments. In *Diapirism and diapirs*, J. Braustein & G. D. O'Brien (eds), 145–61. AAPG Memoir 8.

Moss, A. J. & P. H. Walker 1978. Particle transport by continental water flows in relation to erosion, deposition, soils and human activities. *Sedimentary Geology* **20**, 81–139.

Mulcahy, M. J. 1967. Landscapes, laterites and soils in southwestern Australia. In *Landform studies from Australia and New Guinea*, J. N. Jennings & J. A. Mabbutt (eds), 211–30. Canberra: Australian National University Press.

Neal, E. 1948. *The badger*. London: Collins.

Neilsen, G. A. & F. D. Hole 1964. Earthworms and the development of coprogenous A_1 horizons in forest soils of Wisconsin. *Soil Science Society of America, Proceedings* **28**, 426–30.

Nettleton, W. D., K. W. Flach, B. R. Brasher 1969. Argillic horizons without clay skins. *Soil Science Society of America, Proceedings* **33**, 121–5.

Nettleton, W. D., K. W. Flach, R. E. Nelson 1970. Pedogenic weathering of tonalite in southern California. *Geoderma* **4**, 387–402.

Nikiforoff, C. C. 1949. Weathering and soil evolution. *Soil Science* **67**, 219–30.

Norrish, K. & J. G. Pickering 1983. Clay minerals. In *Soils: an Australian viewpoint*,

CSIRO, 281–308. London: Academic Press.

Northcote, K. H. 1960. *A factual key for the recognition of Australian soils*. Adelaide: Rellim.

Nutting, W. L., M. I. Haverty, J. P. Lafage 1987. Physical and chemical alteration of soil by two subterranean termite species in Sonoran Desert grassland. *Journal of Arid Environments* 12, 233–9.

Nye, P. H. 1954. Some soil forming processes in the humid tropics, I: a field study of a catena in the West African forest. *Journal of Soil Science* 5, 7–21.

— 1955a. Some soil forming processes in the humid tropics, II: the development of the upper slope members of the catena. *Journal of Soil Science* 6, 31–62.

— 1955b. Some soil forming processes in the humid tropics, III: laboratory studies on the development of a typical catena over granite gneiss. *Journal of Soil Science* 6, 63–72.

— 1955c. Some soil forming processes in the humid tropics, IV: the action of the soil fauna. *Journal of Soil Science* 6, 73–83.

Odum, E. P. 1971. *Fundamentals of ecology*, 3rd edn. Philadelphia: Saunders.

Oertel, A. C. 1961. Pedogenesis of some red brown earths based on trace element profiles. *Journal of Soil Science* 12, 242–58.

— 1968. Some observations incompatible with clay illuviation. *Transactions of the 9th International Congress of Soil Science, Adelaide*, vol. 4, 481–8. Sydney: International Society of Soil Science.

— 1974. The development of a typical red brown earth. *Australian Journal of Soil Research* 12, 97–105.

Oertel, A. C. & J. B. Giles 1967. Development of a red brown earth profile. *Australian Journal of Soil Research* 5, 133–47.

Ojanuga, A. G. 1973. Weathering of biotite in soils of a humid tropical climate. *Soil Science Society of America, Proceedings* 37, 644–6.

Ollier, C. D. 1959. A two cycle theory of tropical pedology. *Journal of Soil Science* 10, 137–48.

Pain, C. F. & R. J. Blong 1979. The distribution of tephras in the Papua New Guinea Highlands. *Search* 10, 228–30.

Parham, W. P. 1969. Formation of halloysite from feldspar; low temperature artificial weathering versus natural weathering. *Clays and Clay Minerals* 17, 13–22.

Parizek, E. J. & J. F. Woodruff 1956. Apparent absence of soil creep in the east Georgia Piedmont. *Geological Society of America, Bulletin* 67, 1111–16.

— 1957a. Description and origin of stone layers in soils of the southeastern states. *Journal of Geology* 65, 24–34.

— 1957b. Mass wasting and deformation of trees. *American Journal of Science* 255, 63–70.

Parle, J. N. 1963. A microbial study of earthworm casts. *Journal of General Microbiology* 13, 13–23.

Paton, T. R. 1965. The valley fills of southeastern Queensland. *Australian Journal of Science* 28, 129–30.

— 1974. Origin and terminology for gilgai in Australia. *Geoderma* 11, 221–42.

— 1978. *The formation of soil material*. London: Allen & Unwin.

— 1986. *Perspectives on a dynamic earth*. London: Allen & Unwin.

Paton, T. R., P. B. Mitchell, D. Adamson, R. A. Buchanan, M. D. Fox, G. Bowman 1976. Speed of podzolisation. *Nature* **260**, 601–2.

Patton, P. C., G. Pickup, D. M. Price 1993. Holocene paleofloods of the Ross River, Central Australia. *Quaternary Research* **40**, 1–11.

Pavich, M. J. 1989. Regolith residence time and the concept of surface age of the Piedmont "peneplain". *Geomorphology* **2**, 181–96.

Peters, R. H. 1991. *A critique for ecology*. Cambridge: Cambridge University Press.

Péwé, T. L. 1991. Permafrost. In *The heritage of engineering geology: the first hundred years*, G. A. Kiersch (ed.), 277–298. Geological Society of America, Centennial Special Volume 3.

Phipps, R. L. 1974. The soil creep–curved tree fallacy. *USGS Research Journal* **2**, 371–7.

Pisarski, B. 1978. Comparison of various biomes. In *Production ecology of ants and termites*, M. V. Brian (ed.), 326–42. Cambridge: Cambridge University Press.

Pomeroy, P. E. 1976a. Some effects of mound building termites on soils in Uganda. *Journal of Soil Science* **27**, 377–94.

— 1976b. Studies on a population of large termite mounds in Uganda. *Ecological Entomology* **1**, 49–61.

Price, L. W. 1971. Geomorphic effect of the Arctic ground squirrel in an alpine environment. *Geografiska Annaler* **53A**, 100–6.

Price, R. J. 1973. *Glacial and fluvioglacial landforms*. Edinburgh: Oliver & Boyd.

Prospero, J. M. & T. W. Carlson 1972. Vertical and areal distribution of Saharan dust over the western equatorial North Atlantic Ocean. *Journal of Geophysical Research* **77**, 5255–65.

Prospero, J. M., R. A. Glaccum, R. T. Nees 1981. Atmospheric transport of soil dust from Africa to South America. *Nature* **289**, 570–2.

Pullen, R. A. 1979. Termite hills in Africa; their characteristics and evolution. *Catena* **6**, 267–91.

Pye, K. 1987. *Aeolian dust and dust deposits*. London: Academic Press.

Radwanski, S. A. & C. D. Ollier 1959. A study of an East African catena. *Journal of Soil Science* **10**, 149–68.

Ratcliffe, F. N. & T. Greaves 1940. The subterranean foraging galleries of *Coptotermes lacteus* (Frogg). *CSIRO Journal* **13**, 150–61.

Ratcliffe, F. N., F. J. Gay, T. Greaves 1952. *Australian termites*. Melbourne: CSIRO.

Reeder, C. J. & M. F. Jurgenson 1979. Fire induced water repellency in forest soils of upper Michigan. *Canadian Journal of Forest Research* **9**, 369–73.

Reig, O. A. 1970. Ecological notes on the fossorial octodont rodent *Spalocopus cyanus* (Molina). *Journal of Mammalogy* **51**, 592–601.

Reneau, S. L. & W. E. Dietrich 1990. Depositional history of hollows on steep hillslopes, coastal Oregon and Washington. *National Geographic Research* **6**, 220–30.

Retzer, J. L. 1963. Soil formation and classification of forested mountain lands in the United States. *Soil Science* **96**, 68–74.

Reynolds, R. C. 1971. Clay mineral formation in an alpine environment. *Clays and Clay Minerals* **19**, 361–74.

Robinson, G. W. 1936. Normal erosion as a factor in soil profile development. *Nature* **137**, 950.

Rodin, L. E. & N. I. Bazilevich 1965. *Production and mineral cycling in terrestrial vegetation*. Edinburgh: Oliver & Boyd.

Rogers, L. E. & F. J. Lavigne 1974. Environmental effects of the western harvester ants on the short grass plain ecosystem. *Environmental Entomology* 3, 994–7.

Roose, E. J. 1980. Dynamique actuelle de quelques types de sols en Afrique de l'ouest. *Zeitschrift für Geomorphologie* supplementband 35, 32–39.

Ruhe, R. V. 1956. *Landscape evolution in the High Ituri, Belgian Congo* [INEAC Series Scientifique 66]. Brussels: Government Printer.

— 1959. Stone lines in soils. *Soil Science* 87, 223–31.

— 1984a. Loess-derived soils, Mississippi Valley region, I: soil sedimentation system. *Soil Science Society of America, Journal* 48, 859–63.

— 1984b. Loess-derived soils, Mississippi Valley region, II: soil–climate system. *Soil Science Society of America, Journal* 48, 864–7.

Ruhe, R. V. & J. G. Cady 1955. Latosolic soils of the central African high interior plateaux. *Transactions of the 5th International Congress of Soil Science, Léopoldville*, vol. 4, 401–7. Leopoldville: International Society of Soil Science.

Rutin, J. 1992. Geomorphic activity of rabbits on a coastal sand dune, De Blink dunes, the Netherlands. *Earth Surface Processes and Landforms* 17, 85–94.

Ruxton, B. P. 1958. Weathering and subsurface erosion in granite at the piedmont angle, Balas, Sudan. *Geological Magazine* 95, 353–77.

Ryan, P. J. & J. W. McGarity 1983. The nature and spatial variability of soil properties adjacent to large forest eucalypts. *Soil Science Society of America, Journal* 47, 286–93.

Salem, M. Z. & F. D. Hole 1968. Ant (*Formica exsectoides*) pedoturbation in a forest soil. *Soil Science Society of America, Proceedings* 32, 563–7.

Salick, J., R. Herrera, C. F. Jordon 1983. Termitaria: nutrient patchiness in nutrient deficient rainforest. *Biotropica* 15, 1–7.

Satchell, J. E. 1958. Earthworm biology and soil fertility. *Soils and fertilizers* 21, 209–19.

— 1967. Lumbricidae. In *Soil biology*, A. Burges & F. Raw (eds), 259–322. London: Academic Press.

Saunders, I. & A. Young 1983. Rates of surface processes on slopes, slope retreat and denudation. *Earth Surface Processes and Landforms* 8, 473–501.

Savage, S. M. 1974. Mechanism of fire induced water repellency in soil. *Soil Science Society of America, Proceedings* 38, 652–7.

Schaetzl, R. J., D. L. Johnson, S. F. Burns, T. W. Small 1989. Tree uprooting: review of terminology, process and environmental implications. *Canadian Journal of Forest Research* 19, 1–11.

Schaetzl, R. J. & L. R. Follmer 1990. Longevity of treethrow microtopography: implications for mass wasting. *Geomorphology* 3, 113–23.

Schaetzl, R. J., S. F. Burns, T. W. Small, D. L. Johnson 1990. Tree uprooting: review of types and patterns of soil disturbance. *Physical Geography* 11, 277–91.

Schroo, H. 1964. An inventory of soils and soil suitabilities in West Irian, IIB. *Netherlands Journal of Agricultural Science* 12, 1–26.

Scott, H. C. 1951. The geological work of the mound building ants in western United States. *Journal of Geology* 59, 173–5.

Shachak, M. & Y. Steinberger 1980. An algae–desert snail food chain: energy flow and soil turnover. *Oecologia* 146, 401–11.

Shaler, N. S. 1891. The origin and nature of soils. In *USGS 12th annual report 1890–91*, 213–45.

Sharma, V. N. & M. C. Joshi 1975. Soil excavation by desert gerbil *Meriones hurricane* (Jerdon) in the Shekhawati region of the Rajastan desert. *Annals of Arid Zone* **14**, 268–73.

Sharp, L. A. & W. F. Barr 1960. Preliminary investigations of harvester ants on southern Idaho rangelands. *Journal of Rangeland Management* **13**, 131–4.

Sharpe, C. F. S. 1938. *Landslides and related phenomena*. New York: Columbia University Press.

Sharpley, A. N. & J. K. Syers 1976. Potential role of earthworm casts for the phosphorus enrichment of run-off waters. *Soil Biology and Biochemistry* **8**, 341–46.

— 1977. Seasonal variation in casting activity and in the amounts and release to solution of phosphorous form in earthworm casts. *Soil Biology and Biochemistry* **9**, 227–31.

Sheets, R. G., R. L. Linder, R. B. Dahlgren 1971. Burrow systems of prairie dogs in South Dakota. *Journal of Mammalogy* **52**, 451–2.

Simonson, R. W. 1949. Genesis and classification of red yellow podzolic soils. *Soil Science Society of America, Proceedings* **14**, 316–9.

Slatyer, P. R. O. 1961. Methodology in a water balance study conducted on a desert woodland (*Acacia aneura* F. Muell) community in central Australia. *Arid Zone Research (UNESCO)* **16**, 15–26.

Sprigg, R. C. 1982. Alternating wind cycles of the Quaternary era and their influence on aeolian sedimentation in and around the dune deserts of southeastern Australia. In *Quaternary dust mantles of China, New Zealand and Australia* [Proceedings INQUA Loess Commission Workshop, Canberra, December 1980], R. J. Wasson (ed.), 211–40. Canberra: Australian National University.

Stace, H. C. T., G. D. Hubble, R. Brewer, K. H. Northcote, J. R. Sleeman, M. J. Mulcahy, E. G. Hallsworth 1968. *A handbook of Australian soils*. Adelaide: Rellim.

Statham, I. 1977. *Earth surface sediment transport*. Oxford: Oxford University Press.

Stephens, C. G. 1962. *A manual of Australian soils*. Melbourne: CSIRO.

Stoops, G. 1964. Application of some pedological methods to the analysis of termite mounds. In *Etudes sur les termites africains*, A. Bouillon (ed.), 379–98. Leopoldville: Leopoldville University Press.

— 1967. Micromorphology of some characteristic soils of the Lower Congo. *Pédologie* **18**, 110–49.

Swindale, L. D. 1975. The crystallography of minerals of the kaolin group. In *Soil components*, vol. 2: *inorganic components*, J. E. Giesking (ed.), 121–54. New York: Springer.

Sys, C. 1960. *Notice explicative de la carte des sols du Congo Belge et du Ruanda–Urundi*. Brussels: Institut National pour l'Etude Agronomique du Congo Belge (INEAC).

Talbot, M. 1953. Ants of an old field community on the Edwin S. George Reserve, Livingston County, Michigan. *Contributions of the Laboratory of Vertebrate Biology, University of Michigan* **69**, 1–9.

Taylor, N. H. 1949. Soil survey and classification in New Zealand. *Proceedings of the 7th Pacific Science Congress*, vol. 6, 103–13. Wellington: Whitcombe & Tombs.

Tazaki, K. 1986. Observations of primitive clay precursors during microcline weathering. *Contributions to Mineralogy and Petrology* **92**, 86–8.

Tazaki, K. & W. S. Fyfe 1987. Primitive clay precursors formed on feldspar. *Canadian Journal of Earth Sciences* **24**, 506–27.

Terwilliger, V. J. & L. J. Waldron 1990. Assessing the contributions of roots to the strength of undisturbed slip prone soils. *Catena* **17**, 151–62.

Thompson, C. H. & T. R. Paton 1980. *Texture differentiation in soils on hillslopes of southeastern Queensland*. Division of Soils Report 53, CSIRO, Adelaide.

Thorn, C. E. 1978. A preliminary assessment of the geomorphic role of pocket gophers in the alpine zone of the Colorado Front Range. *Geografiska Annaler* **60A**, 181–8.

Thorp, J. 1949. Effect of certain animals that live in soils. *Science Monthly* **68**, 180–91.

Thorp, J. & G. D. Smith 1949. Higher categories of soil classification: order, suborder and great soil groups. *Soil Science* **67**, 117–26.

Townshend, J. R. G. 1970. Geology, slope form and slope process and their relation to the occurrence of laterite. Part V of the Symposium on the Royal Society/Royal Geographical Society expedition to northeastern Mato Grosso, Brazil. *Geographical Journal* **136**, 392–99.

Trapnell, C. G. 1943. *The soils, vegetation and agriculture of northeastern Rhodesia*. Lusaka: Government Printer.

Trapnell, C. G. & J. N. Clothier 1937. *The soils, vegetation and agricultural systems of northwestern Rhodesia*. Lusaka: Government Printer.

USDA (United States Department of Agriculture) 1938. *Soils and men* [yearbook of agriculture], 948–92. Washington DC: US Government Printing Office.

— 1951. *Soil survey manual* [Handbook 18]. Washington DC: US Government Printing Office.

— 1960. *Soil classification, a comprehensive system* (7th approximation). Washington DC: US Government Printing Office.

— 1975. *Soil taxonomy* [Handbook 436]. Washington DC: US Government Printing Office.

van Hise, C. R. 1904. *A treatise on metamorphism*. USGS Monograph 47.

van Wambeke, A. 1959. Le rapport limon/argile, mesure approximative, du stade d'alteration des materiaux originals des sols tropicaux. *Proceedings of the 3rd Inter-African Soils Conference, Dalaba*, vol. 1, 161–7. Lagos: Inter-African Pedological Service.

van Zon, H. J. M. 1980. The transport of leaves and sediment over a forest floor. *Catena* **7**, 97–110.

Vedy, J. C. & S. Bruckert 1979. Soil solution: composition and pedogenic significance. In *Constituents and properties of soils*, M. Bonneau & B. Souchier (eds), 184–213. London: Academic Press.

Viereck, L. G., B. J. Griffin, H-U. Schmincke, R. G. Pritchard 1982. Volcaniclastic rocks of the Reydarfjordur drill hole, eastern Iceland, 2: alteration. *Journal of Geophysical Research* **87**(B8), 6459–76.

Voslamber, B. & A. W. L. Veen 1985. Digging by badgers and rabbits on some wooded slopes in Belgium. *Earth Surface Processes and Landforms* **10**, 79–82.

Wada, K. 1979. Structural formulas of allophanes. *Proceedings of the 6th International Clay Conference, Oxford*, M. M. Mortland & V. C. Farmer (eds), 537–45. Amsterdam: Elsevier.

— 1981. Amorphous clay minerals – chemical composition, crystalline state, synthesis and surface properties. *Proceedings of the 7th International Clay Conference, Bologna*, H. van Olphen & F. Veniale (eds), 385–98. Amsterdam: Elsevier.

— 1985. The distinctive properties of andosols. *Advances in Soil Science* 2, 173–229.

Wada, K. & I. Matsubara 1968. Differential formation of allophane, "imogolite" and gibbsite in the Kitakami pumice bed. *Proceedings of the 9th International Congress of Soil Science, Adelaide*, vol. 3, 123–30. Sydney: International Society of Soil Science.

Walker, E. P. (ed.) 1968. *Mammals of the world*, 2nd edn. Baltimore: The Johns Hopkins University Press.

Walker, P. H. & D. J. Chittleborough 1986. Development of particle size distributions in some alfisols of southeastern Australia. *Soil Science Society of America, Journal* 50, 394–400.

Walker, P. H. & J. Hutka 1979. Size characteristics of soils and sediments with special reference to clay fractions. *Australian Journal of Soil Research* 17, 383–404.

Walker, P. H., P. I. A. Kinnell, P. Green 1978. Transport of a non-cohesive sandy mixture in rainfall and runoff experiments. *Soil Science Society of America, Journal* 42, 793–801.

Waloff, N. & R. E. Blackith 1962. The growth and distribution of the mounds of *Lasius flavus* (Fabricius) (Hym: Formicidae) in Silwood Park, Berkshire. *Journal of Animal Ecology* 31, 421–37.

Wasson, R. J. 1982. Contribution of dust to Quaternary valley fills at Belarabon, western NSW. In *Quaternary dust mantles of China, New Zealand and Australia* [Proceedings INQUA Loess Commission Workshop, Canberra, December 1980], R. J. Wasson (ed.), 191–6. Canberra: Australian National University.

Watanabe, H. 1975. On the amount of cast production by the megascolecid earthworm *Pheretima hupiensis*. *Pedobiologia* 15, 20–8.

Watanabe, H. & S. Ruaysoongnern 1984. Cast production by the megascolecid earthworm *Pheretima* sp. in northeastern Thailand. *Pedobiologia* 26, 37–44.

Watson, J. A. L. & F. J. Gay 1970. The role of grass eating termites in the degradation of amulga ecosystem. *Search* 1, 43.

Watson, J. P. 1964. A soil catena on granite in Southern Rhodesia. *Journal of Soil Science* 15, 239–57.

Weaver, C. E. 1989. *Clay, muds and shales* [Developments in Sedimentology 44]. New York: Elsevier.

Webley, D. M., M. E. K. Henderson, I. F. Taylor 1963. The microbiology of rocks and weathered stones. *Journal of Soil Science* 14, 102–112.

Webster, R. 1965. A catena of soils on the Northern Rhodesian Plateau. *Journal of Soil Science* 16, 31–43.

Wells, S. G. & J. C. Dohrenwend 1985. Relict sheetflood bed forms on late Quaternary alluvial-fan surfaces in the southwestern United States. *Geology* 13, 512–16.

Wheeler, W. M. 1910. *Ants: their structure development and behaviour*. New York: Columbia University Press.

White, L. P. 1970. Brousse tigrée patterns in southern Niger. *Journal of Ecology* 58, 549–53.

— 1971. Vegetation stripes on sheet-wash surfaces. *Journal of Ecology* 59, 615–22.

White, R. S. & D. P. McKenzie 1989. Volcanism at rifts. *Scientific American* 261, 62–71.

Wiken, E. B., K. Broersma, L. M. Lavkulich, L. Farstad 1976. Biosynthetic alteration in a British Columbia soil by ants (*Formica fusca*, Linné). *Soil Science Society of America, Proceedings* **40**, 422–6.

Williams, M. A. J. 1968. Termites and soil development near Brock's Creek, Northern Territory. *Australian Journal of Science* **31**, 153–4.

Williams, W. D. 1974. Freshwater crustacea. In *Biogeography and ecology in Tasmania*, W. D. Williams (ed.), 63–112.

Wilshire, H. G. 1958. Alteration of olivine and orthopyroxene in basic lavas and shallow intrusions. *American Mineralogist* **43**, 120–47.

Wilson, L. 1972. Explosive volcanic eruptions, II: the atmospheric trajectories of pyroclasts. *Royal Astronomical Society, Geophysical Journal* **45**, 543–56.

Wilson, M. J. 1969. A gibbsitic soil derived from the weathering of an ultrabasic rock on the island of Rhum. *Scottish Geology* **5**, 81–9.

Wilson, M. J. & D. Jones 1983. Lichen weathering of minerals: implications for pedogenesis. In *Residual deposits: surface related weathering processes and materials*, R. C. L. Wilson (ed.), 5–12. Oxford: Blackwell Scientific.

Wilson, M. J., D. Jones, W. J. McHardy 1981. The weathering of serpentinite by *Leconaraatra. Lichenologist* **13**, 167–76.

Wood, A. W. 1987. The humic brown soils of the Papua New Guinea highlands: a reinterpretation. *Mountain Research and Development* **7**, 145–56.

Wood, T. G. & K. E. Lee 1971. Abundance of mounds and competition among colonies of some Australian termite species. *Pedobiologia* **11**, 341–66.

Worrall, G. A. 1959. The butana grass patterns. *Journal of Soil Science* **10**, 34–53.

Yaalon, D. H. & J. Dan 1974. Accumulation and distribution of loess-derived deposits in the semi-desert and desert fringe areas of Israel. *Zeitschrift für Geomorphologie* supplementband **20**, 91–105.

Yair, A. & J. Rutin 1981. Some aspects of regional variation in the amount of available sediment produced by isopods and porcupines, Northern Negev, Israel. *Earth Surface Processes and Landforms* **6**, 221–34.

Young, A. & I. Saunders 1986. Rates of surface processes and denudation. In *Hillslope processes*, A. D. Abrahams (ed.), 3–27. Boston: Allen & Unwin.

Young, A. & I. Stephen 1965. Rock weathering and soil formation on high altitude plateaux of Malawi. *Journal of Soil Science* **16**, 322–33.

Young, A. R. M. 1983. Patterned ground on the Woronora Plateau. In *Aspects of Australian sandstone landscapes* [Australian and New Zealand Geomorphology Research Group Special Publication 1], R. W. Young & G. C. Nanson (eds), 48–53.

Zlotin, R. I. & K. S. Khodashova 1980. *The role of animals in biological cycling of forest-steppe ecosystems*. Stroudsburg, Pennsylvania: Dowden, Hutchinson & Ross.

Author index

Subject index